The Migrant Farmer
in the History of
the Cape Colony
1657–1842

1885 drawing by Heinrich Egersdorfer (1853–1915)
made for *Illustrated London News*

The Migrant Farmer
in the History of
the Cape Colony
1657–1842

P. J. van der Merwe

Translated by
Roger B. Beck

OHIO UNIVERSITY PRESS
ATHENS

All other rights to *Die Trekboer in die Geskiedenis van die Kaåpkolonie, 1657–1842*
held by Mrs. P. J. Van der Merwe
99 98 97 96 95   5 4 3 2 1

Ohio University Press books are printed on acid-free paper ∞

Library of Congress Cataloging-in-Publication Data

Van der Merwe, P. J. (Petrus Johannes), 1912–1979.
   [Trekboer in die geskiendenis van die Kaapcolonie, 1657–1842.
English]
   The migrant farmer in the history of the Cape colony, 1657–1842/
P.J. van der Merwe ; translated by Roger B. Beck.
     p.   cm.
   Includes bibliographical references and index.
   ISBN 0-8214-1090-3
   1. Cape of Good Hope (South Africa)—History—To 1795. 2. Cape of
Good Hope (South Africa)—History—1795–1872. 3. Migrant
agricultural laborers—South Africa—Cape of Good Hope—History.
I. Beck, Roger B.
DT1813.V3613 1995
968.7—dc20
                                  94-21649
                                     CIP

Contents

Translator's Introduction

Petrus Johannes van der Merwe wrote three of the most significant books on the history of South Africa before he was thirty-five years old. The three volumes of his trilogy—*Die Trekboer in die Geskiedenis van die Kaapkolonie (1657–1842)*, *Die Noordwaartse Beweging van die Boere voor die Groot Trek* (the English title will be *The Northward Movement of the Farmers before the Great Trek*), and *Trek: Studies oor die Mobiliteit van die Pioniersbevolking aan die Kaap* (*Trek: Studies of the Mobility of the Pioneer Population at the Cape*)—have become classics that no student of Cape colonial history of the seventeenth, eighteenth, or nineteenth century can ignore. Today van der Merwe is recognized as one of the finest of South African historians, even though his writings have been inaccessible to all but the relatively small number of people who can read Afrikaans. Hence an English translation is needed that will make his works available to a world-wide audience.

Born in Griquatown in the northern Cape in 1912, van der Merwe received both his B.A. (1931) and M.A. (1933) degrees from the University of Stellenbosch. His M.A. thesis, "Die Geskiedenis van die Trekboer onder die Oos-Indiese Companje" ("The History of the Migrant Farmer under the East India Company"), was written under the direction of professor Willem Blommaert, to whom this present volume is dedicated. In 1934 van der Merwe went to study at the University of Leiden, where he worked on his doctoral degree under the famous Dutch cultural historian J. H. Huizinga. During the three years he lived in Europe, van der Merwe traveled widely, carrying out research in the state archives in Munich and Berlin, in the Public Record Office and London Missionary Society archives in London, and in the state archives in the Hague. He defended his doctoral thesis, "Die Noordwaartse Beweging van die Boere voor die Groot Trek, 1770–1842," in September 1937 and it was published that year in the Hague.

Van der Merwe returned to South Africa at the end of 1937 and at the beginning of 1938 renewed his long association with the University of Stellenbosch, now as a lecturer in the history department. He also published the present volume in 1938. Van der Merwe was thirty-three years old when *Trek* was published by Nasionale Pers in Cape Town in 1945. Two minor works also appeared during this early period: *Die Kafferoorlog van 1793* (*The Frontier War of 1793*) in 1940 and *Pioniers van die Dorsland* (*Pioneers of the Dorsland*) in 1941. He was awarded a prize in 1948 by the South African Academy of Arts and Sciences for his original contributions to South African historical writing. Van der Merwe was to publish only one other major work, *Nog verder noord, die Potgieter-kommissie se besoek aan die gebied van die teenswoordige Suid-Rhodesië, 1836* (*Still Farther North, the Potgieter Patrol's Visit to the Region of Present-Day Southern Rhodesia, 1836*) (1962), and a brief style manual for historical publications, *Bronnelys en voetnote* (1972) (published in English as *Source List and Footnotes*). In 1987 an edited version of a manuscript found after his death in 1979 was published in the Archives Yearbook series with the title, *Die Matebeles en die Voortrekkers* (*The Matebeles and the Voortrekkers*).

P. J. van der Merwe was unique among Afrikaner historians in that he focused not on the single event known as the Great Trek, but on the greater, nearly three-hundred-year-long migration of peoples of Dutch, French, and German descent out from the victualing station at Cape Town after their arrival there in 1652. In the process he pioneered new directions in historical writing decades before they became fashionable among other South African historians. Van der Merwe was less interested in politics than in the social, cultural, economic, and religious lives of his subjects. No doubt influenced by Huizinga, his mentor in Leiden, van der Merwe asked questions about such daily concerns as work, food, property owning, private and public worship, leisure activities, fashions, the environment, and the farmers' relations with their neighbors, both white and black.

Readers may be offended by some of the racial references and glorifications of the Afrikaner past contained in this book. Neither the publisher nor the translator share these views, and the work is not being reissued in an English edition in order to advance them in any way. The purpose of this English edition is to make a classic in South African historical writing accessible to a world readership and to allow more scholars the opportunity to use, or challenge, the wealth of information and pioneering analysis contained in it.

While certainly a man of his time and of his Afrikaner people, P. J. van der Merwe probably brought less of an ideological bent to his work than did any of his contemporaries, and many since. He was meticulous in his research, visiting libraries and archives throughout Europe and southern Africa and traveling thousands of miles by car in order to visit the places about which he wrote and to interview the peoples who still lived there. The emphasis he placed on supporting his conclusions and theories with documentary evidence often led him into conflict with contemporary myth or the accepted ideologically or politically motivated "truth" of the time. This conflict is most evident in the present volume when he challenges the popular conclusion of George McCall Theal that hostilities on the eastern Cape frontier between black and white were caused by Xhosa cattle stealing and raids. While paying his respects to an icon of South African historiography, the then twenty-six-year-old historian systematically employs his mass of archival data to show that Theal's conclusions were not supported by the evidence.

Van der Merwe was writing a history of his people and naturally concentrated on their lives and activities. The reader will find little here about the daily lives of the indigenous peoples of southern Africa. But when the two sides met, van der Merwe was concerned to show exactly what happened according to the evidence available. Ken Smith writes,

> He was at pains not to judge either the *trekboers* or the San for their conduct on the frontiers, but to see their actions in terms of the situation existing at the time. Thus, the Boers, with their notions of private ownership of goods, could not be blamed for shooting San who stole their cattle any more than the San, who had no conceptions of private ownership, could be blamed for taking cattle and sheep when they were hungry.

As Smith rightly points out, this approach would be seriously challenged today, but in the context of the late 1930s and early 1940s, van der Merwe was framing questions that Afrikaner historians as a whole were hardly beginning to ask in the 1970s. His broad approach and attempt to see both sides of the question, his probing of the social aspects of life among the Boers in the interior, was unique.[1]

Much research has been conducted and much has been written since van der Merwe's works first appeared, but after fifty years these three books still remain essential contributions to the early history of the Cape. The best single source containing the most recent data and theory concerning the time frame and many of the same themes as van der Merwe's studies is *The Shaping of South African Society, 1652–1840*, coedited by Richard Elphick and Hermann Giliomee.[2] The interested scholar will do well to begin here when going in search of bibliography, current research data, and historiographical debate. But even in this modern study, the value of van der Merwe's work is still evident. William M. Freund, for example, notes that "the most important source on the land question at this time remains P. J. van der Merwe, *Die Trekboer. . . .*"[3] Hermann Giliomee writes that "the discussion of the causes of trekboer expansion has been dominated by P. J. van der Merwe. . . ."[4], and Robert Ross, in reference to colonial settlement, particularly north of the Orange River, states that "the standard work on this is still P. J. van der Merwe, *Die Noordwaartse Beweging. . . .*"[5] Not only will this English edition make these important contributions accessible to a worldwide audience, but hopefully it will cause more scholars to focus on this important period in South African history.

Space does not permit a fuller account of van der Merwe's life and work here. Readers who are interested may find an excellent analysis of his contributions to, and place in, South African historiography in Ken Smith, *The Changing Past: Trends in South African Historical Writing*.[6] The most extensive biography of van der Merwe, regrettably available only in Afrikaans, is "P. J. van der Merwe: Ondersoeker van die Afrikaner se landelike pioniersgeskiedenis," ("P. J. van der Merwe: Researcher of the Rural Afrikaner's Pioneer History," in *Wie en wat is die Afrikaner? (Who and What is the Afrikaner?*).[7]

Every book presents its own unique problems to the translator. One of the most challenging problems in the present work was van der Merwe's intermixing of several different languages. Although the majority of the text is in Afrikaans, there are also substantial passages in seventeenth and eighteenth century Dutch and German as well as English. Although some of the works from which these quotations are taken are now available in English editions, I have chosen to offer my own translation of the entire book and not to translate some here and borrow from others there. I hope this will make reading of the work more consistent throughout.

Another major problem is the one that faces all scholars writing about South Africa: that of appropriate terminology. I have changed the pejorative terms *kaffir* and *Hottentot* to *Xhosa* and *Khoikhoi* except when used in a quotation. The debate over the use of *San* in place of *Bushman* continues, but most scholars today seem to be leaning toward Bushman and that is the term I use.

Some South African readers may find fault with me for translating such common terms as *boer, trekboer, kraal,* and *karosse.* Although those terms are widely understood in South Africa, they would not be in the United States or among English readers in, say, Nigeria or India. As one example, *trek* and *boer* are recognized to some extent internationally, but in combination with other words—such as *trekkos, trekpad, trekgees, boerwoning, veeboer* and *boerdery*—they are not as familiar to non-South African readers and had to be rendered in English. To be consistent I have translated every word except *landdrost* and *heemraad* (pl. *heemraden*), special political terms used to describe local district officials. The landdrost was the chief magistrate of a district, and a heemraad belonged to the board of heemraden that advised and assisted the landdrost with administration of the district. The local council was also collectively known as the heemraad. The territory over which they had jurisdiction, as well as the landdrost's residence, was known as the *drostdy.*

Finally, a note about the footnotes and bibliography. The Cape Archives, and to a lesser extent the Dutch national archives in the Hague, have changed their systems of cataloguing archival materials since van der Merwe conducted his research in the 1930s. In order to make this edition more useful to the modern scholar I have updated all the call numbers in the footnotes for the archives in Cape Town and the Hague. The relevant call numbers for docu-

ments in the state archives in Berlin have remained unchanged.
While making these changes I became aware of a substantial number of errors in the footnotes themselves and in the bibliography.
For the footnotes, those errors included inaccurate dates, volume
and page numbers, titles, and inconsistent citations. I have checked
each of the footnotes and made corrections where necessary. The
entries for a number of books in the bibliography were incomplete,
and several books and archival sources were left out altogether. I
have corrected those errors and omissions. In the footnotes I have
translated most of the titles for the archival series, but decided to
leave the old call numbers and original Afrikaans, Dutch, or German descriptors as van der Merwe had them in the bibliography
because these include all the volumes consulted by him in the archives, and because those descriptors are still used by the archives
for many series. The interested researcher may find the new numbers by referring to the footnotes or by using the indexes available
at the different archives.

NOTES

1. Ken Smith, *The Changing Past: Trends in South African Historical
Writing* (Athens: Ohio University Press, 1988), p. 78.
2. 2d ed. (Middletown, Conn.: Wesleyan University Press, 1989).
3. William M. Freund, "The Cape Under the Transitional Govern-
ments, 1795–1814," in *The Shaping of South African Society*, ed. Richard
Elphick and Hermann Giliomee (Middletown, Conn.: Wesleyan University Press, 1989), p. 355, fn. 20.
4. "The Eastern Frontier, 1770–1812," in *Shaping*, p. 424, fn. ‡
5. "The Cape of Good Hope and the World Economy, 1652–1835,"
in *Shaping*, p. 280, fn. 147.
6. Pp. 76–80.
7. F. A. van Jaarsveld, *Wie en wat is die Afrikaner?* (Cape Town:
Tafelberg, 1981), pp. 129–91.

Acknowledgments

My work on this project was funded in part by three faculty research grants that I received from Eastern Illinois University. My special thanks go to Nola Heyns, who read my initial very rough draft and encouraged me to continue, and to Edwin Hees, Department of English, University of Stellenbosch, who did the final reading and editing of the manuscript. Robert Ross, University of Leiden, was my source of last resort for the Dutch passages, for several arcane references in the bibliography, and for the cataloguing system in the Hague archives. I also had help with the Dutch and Afrikaans from Robert Shell and Leonard Guelke, and with the Dutch and German from Jim Riley, William Shetter, and Wolfgang Schlauch. The staff at the Cape Archives was invaluable in helping me track down footnote references to their holdings. I spent a month in the Cory Library, Rhodes University, looking up references to published works and owe a special debt of gratitude to Sandy Rowoldt, chief archivist at Cory. Peter Jörg Becker, Staatsbibliothek zu Berlin-Preußsischer Kulturbesitz, checked the bibliographical information for me in the state archives in Berlin. I also am deeply indebted to the Head of Special Collections, Hanna Botha, and to the Africana librarian, Daleen van der Riet, at the J. S. Gericke Library, University of Stellenbosch for their kind assistance with the footnotes and bibliography. Johannes Marais, an attorney in Stellenbosch, very kindly provided me with secretarial services through his law firm, and transportation and housing during my stay there. Susie Newton-King graciously allowed me to stay at her house in Cape Town while I worked on the footnotes. The two diagrams in the original edition have been redrawn to make them clearer, and I had a new map designed that is clearer and shows more of the places mentioned in the book. My thanks to Steve DiNaso and Vince Gutowski of the Eastern Illinois University Geography Department for their excellent work.

This translation would not have been possible without the permission and support of Mrs. M. E. van der Merwe, Dr. van der Merwe's widow. Our conversations have allowed me to gain a better understanding of her late husband's life and his attitude and approach to his work. I cannot thank her enough for her encouragement and only hope she finds this translation worthy of the high standards of scholarship set by Professor van der Merwe.

Preface

This study forms the first part of a projected trilogy about the pioneer history of the Cape Colony, of which the third part—*Die Noordwaartse Beweging van die Boere voor die Groot Trek* [*The Northward Movement of the Farmers before the Great Trek*]—appeared in 1937. The title of the second part will be *Die Mobiliteit van die Pioniersbevolking aan de Kaap* [*The Mobility of the Pioneers at the Cape*]. I hope in that work to answer most of the questions that the reader may still pose, after having read the two published volumes.

Die Trekboer in die Geskiedenis van die Kaapkolonie must not be regarded and judged as a fully rounded monograph about one or another clearly defined subject. The individual chapters were written and intended as separate studies about various subjects. These studies, however, all deal with the seminomadic frontier farmer— *die trekboere* [the migrant farmers]—who led the natural expansion of the colony into the interior.

I would also like to take this opportunity to express my appreciation for the help I have received from a number of people. In the first place I owe many thanks to two Dutchmen: Mr. Swellengrebel, who was kind enough to place his family archives at my disposal, and Mr. Hallema, through whose good offices I obtained the documents for examination. From Professor Doctor W. M. R. Malherbe I received valuable help in connection with the second chapter; Mr. P. J. Venter, Cape Town archivist, for his unselfish assistance; Mr. Eric Stockenström drew the map and Mr. P. J. Cilié very kindly read through the manuscript. I offer them all my heartfelt thanks. I extend a word of special thanks to the National Research Council of the Union of South Africa, from whom I received financial help in connection with the publication of this work.

Finally I acknowledge with sadness the premature death of my professor [W. Blommaert], to whose memory I dedicate this work with gratitude and thankfulness. Through his encouragement I began the study of our pioneer history and under his inspiring leadership I conducted my first research work. May this book contribute to keeping his memory alive by those whom he so un-selfishly served.

P. J. v.d. M.

The University,
Stellenbosch,
September 1938

In memory of
Prof. Dr. W. Blommaert

I

The Evolution of the Migrant Farmer*

THE ORIGIN OF LIVESTOCK FARMING

The First Free Burghers

When the Company announced to the fort its readiness to establish free colonists at the Cape, a few soldiers quickly indicated that they were prepared to accept the offer. They had great expectations about the future and optimistically discussed their farming plans. They would establish two "colonies." The first, named "Harmans Colony . . . would concentrate principally on wheat farming," and the second, "Stevens Colony . . . would produce, together with grain, tobacco, a variety of vegetables and fruits. Both settlements are to rear cattle, pigs, geese, ducks, chickens, etc., from which they may derive a profit, yet the Company's principal purpose for these settlements remains the promotion of agriculture first and the rest at a later time."[1] The commander consented to their plans but at the same time warned them that they must become *farmers*. They would not be allowed to indulge in secondary activities that could distract them from farming, "but should concentrate on agriculture, cattle breeding, etc."[2]

Before long, however, the so-called free burghers were to experience a great many disillusionments. Having left the Company's service in order to be free and to look after their own welfare, they now found the commander was trying to keep them under his thumb, and was severely restricting the sources of income on which

---

*This chapter's title is taken from Prof. Leo Fouché's published lecture that raised interest in this interesting subject. See Leo Fouché, *Die Evolutie van die Trekboer: Lesing gehou voor die Christelike Jongelieden Vereniging* (Pretoria: Volkstem, 1909).

they had originally reckoned. They had to sow and plant what he prescribed. In spite of repeated failures with rice cultivation, they were still compelled to sow a certain amount each year. They were forbidden to plant vegetables because this might distract them from wheat farming—the work for which they had been settled on the land. Tobacco growing was not allowed because it would ruin the market and would lead, moreover, to illicit trading with the Khoikhoi.[3] The free burghers' greatest grievance, however, was the abolition of Commissioner Ryckloff van Goens's cattle trade concession. The former soldiers had secretly hoped to enrich themselves quickly through the profitable cattle barter with the Khoikhoi. This was evidently one of the most attractive reasons for becoming a free burgher. "We obtained our freedom," they complained bitterly in 1658, "that we might freely trade in cattle, though not excluding other articles, with the inhabitants called Hottentots, yet now that we are free and have put in much arduous labor upon the land, that trade is forbidden to us."[4]

The government was thus determined to make good farmers of the new settlers, and no choice remained for them but to set to work diligently. Unfortunately, the opportunity to make a good living from wheat farming was clearly diminished by the fact that the free burghers were denied the right to a free market. They could sell their produce only to the Company, and it alone would set the prices.[5]

In 1658 the free burghers optimistically began to sow wheat before a definite purchase price had been settled upon. When the crop ripened, however, they became impatient and wanted to know what they were going to receive for it. They questioned the commander repeatedly, but his responses were evasive. He had received the desired information long before,[6] but the prices were about half what the free burghers had expected, and even he thought they had been fixed too low. So as not to discourage the farmers, he left them in ignorance while continuing to negotiate with the Lords XVII.

The farmers were unwilling to be kept in the dark for too long, however, and on 23 December 1658 they appeared before the commander with their first petition. They demanded that he make known the purchase prices before they would thresh their already harvested grain. "Set us a price soon," drummed the impudent request, "because we will farm no land as long as we do not know the price, for we do not wish to be Company slaves." Moreover, they refused to sell their wheat for under ten guilders a muid. If they could not get that, "they did not intend to farm even one more foot of land."[7]

Van Riebeeck, the commander governor, tactfully succeeded in assuaging the rebellious settlers, with whom he sympathized. He realized higher grain prices were necessary, not only for the "common good and maintenance of the free citizens," but also to advance the Company's own interests. It was always the Company's stated aim to promote agriculture, from which "not much could be expected without the promise of a fair price."[8] The commander could do nothing without instructions from the Lords XVII, however, and the farmers had to be satisfied with less than seven guilders a muid—a price guaranteed not to stimulate much enthusiasm among them for wheat farming.

This first confrontation between the farmers and government was just the beginning of a long dispute over higher grain prices. And when the government, which tried to monopolize the market, would not pay subsistence prices, the farmers sought compensation in the smuggling trade with passing ships, independent bakers, and other free colonists.[9] In this way, and in spite of numerous laws forbidding it, the free burghers often succeeded in obtaining relatively high prices for their grain as long as the demand exceeded the supply. (In 1666, for example, commander Wagenaer told of a colonist who committed the scandalous act of selling a sack of wheat for twelve guilders.(!)[10] But the colonists' welfare depended primarily on the price the Company paid them for their grain, and the Company held it so low that the free burghers remained poor and wheat farming was not able to develop successfully.[11]

Consequently, the colonists appeared before the Council of Policy again in 1663 with a petition. They complained they could not make any profits from agriculture, and asked the commander to raise the grain price.[12] The Cape authorities appreciated the grounds for their complaints and the reasonableness of their petition and pleaded their case before the visiting commissioner. He considered their petition favorably, and toward the end of the year grain prices were raised a little. But this did not help much because production costs associated with agriculture at the Cape were very high.

The colonists did receive their land free, but it had become very difficult to find suitable laborers. Initially the farmers used mainly Company soldiers, "loaned" to them by the government with a stoppage of wages. These soldiers were not really very suitable workers. More often than not they were lazy and often ran away at times when the farmers most needed them.[13] In addition, military labor was very costly. First, the farmers had to provide for these soldiers, and victuals were expensive. Furthermore, they had to pay the soldiers relatively high wages. It was not unusual for a farmer to pay out wages of two hundred to four hundred guilders a year.[14] And this money had to be paid out in cash—a very scarce item among the free burghers, the majority of whom had commenced farming with nothing.

But military labor was not only inferior and expensive; sometimes it was difficult even to find enough workers. More than once a labor shortage prevented the free burghers from harvesting and threshing the crops standing ripe in the fields.[15] Native labor played a relatively minor role at first. The nomad Khoikhoi who occasionally visited the fort were never good agricultural laborers. Especially in the beginning they were not willing to abandon their lifestyles and go to work for the farmers. The labor question was resolved satisfactorily later by importing slaves, but slave labor was terribly expensive.

Wheat farming was also a very uncertain business because of the fickle nature of the Cape climate. The vicious southeast wind in particular often blew without letup during harvest time, repeatedly causing the farmers severe losses. This wind often pulled down the ripe ears of wheat just before they were to be reaped, thus almost completely destroying a promising harvest. Sometimes the farmers barely recovered their seed wheat.[16] Up to the time of Simon van der Stel the government tried in vain to find a place that was suitable for grain farming and free from this scourge as well. There was no such place on the Peninsula and even in Hottentots Holland—although more protected than the areas in and around Cape Town—the southeast wind repeatedly caused great damage.[17] It was only when Simon van der Stel made his famous journey over Helshoogte that he discovered a beautiful valley that was actually completely free from this destructive wind. In this pretty setting a new colony was established shortly thereafter, which was called Stellenbosch.

Crop failures, added to low grain prices and high production costs, hampered agricultural development at the Cape. The colonists could not make a decent living from agriculture. Under normal conditions the yield from the harvest barely covered the high production costs.[18] The colonists often had so little left over they could not support their wives and children. One bad year meant disaster. From early morning till late evening the free burghers turned up the ground like moles—many times alone, if they could not afford white servants—but despite these efforts they sank deeper into debt.[19] Circumstances over which they had no control kept even the hardest working among them in a state of pitiable poverty.[20] The situation became so critical in 1664 that the Council of Policy decided that the income from the Sunday collections and fines, amounting at that time to 1,268 guilders and 16 stuiwers, would be distributed for the benefit of the poor colonists, who were "burdened with naked children, and from simple poverty must sleep each night beside the livestock in the stable on a little straw and the barren earth."[21]

About 1670 the Cape government began to fear that wheat farming would significantly decline. So as not to frustrate their own purposes, the price for wheat was raised to ten guilders per muid.[22] This brought some relief to be sure, but even at this price the farmers could not produce at a profit. In 1677 the Burgher Senate and "the best and most capable or diligent and sober citizens" calculated that wheat production costs amounted to eleven guilders per muid, before it was even threshed and put away in the storehouse.[23] This calculation was perhaps a little high, since it was not made disinterestedly, but there remains little doubt that the natural price for wheat at this time was higher than the market price.[24] Under these conditions even the very best producers had to live thriftily, even when all was going well, and even then a bad harvest meant debt or poverty.

Agriculture Becomes Unpopular

Inevitably the free burghers began to lose interest in their work and evince a growing aversion to farming because of the poor prospects offered by agriculture. The government was often only too ready to look for the cause of this in the colonists' laziness. In 1662, for example, Wagenaer reported that some free burghers who had taken good care of their farms had made reasonable progress, "mainly with the breeding of various types of large and small livestock."[25] But there were only six or eight hard workers among them. The rest "are drunken, lazy, churlish, and insolent fellows,"[26] who had not given proper attention to agriculture, but had entrusted everything to the care of servants, while they occupied themselves with forbidden practices. He even suggested that a number of free burghers, together with their servants, ought to be deported or obliged to return to Company service.[27] Deportation was undesirable, however, because the majority of free burghers were up to their ears in debt to the Company. If they were forced to leave the colony, the Company would lose the money!

Wagenaer's successors also repeatedly accused the free burghers of laziness. But just like him, they recognized as well that even the hardest-working colonists had made no progress despite all their laborious toil on the land.[28] This inconsistency must be explained by the fact that the Company did not regard the free burghers as men who were free to work out their own destinies. From the Company's point of view, the free burghers were there to produce grain at a low price. This would eliminate the need to import rice from India and thereby save costly space on the heavily laden, homeward-bound trading ships. The free burghers, however, viewed their farming ventures from a different perspective, and rightly so. They regarded wheat farming as a means of earning a livelihood, and they would apply themselves diligently only if they could earn a profit from it. If this were impossible, they were ready to switch to something that was profitable.

As wheat farming was not economically attractive, it was not long before the free burghers began to neglect it. The commander did his best to encourage them to produce grain. He gave them free land, and oxen with which to work it, but to little avail. Many farmers allowed their freehold farms to lie uncultivated because the prospects for earning a profit—the most effective incentive for effort—were lacking. The commander discovered with regret that the free burghers were only too ready to barter away their oxen in order to obtain money to begin something else.[29] An increase in the grain price would certainly have stimulated the colonists to till their land, but a price increase was contrary to the Company's shortsighted policy that aimed at direct profits—not the development of the colony.

Despite repeated and earnest admonitions to cultivate their land, a large number of the supposed grain producers did not even harvest wheat for their own bread. They pestered the commander repeatedly for rice. This placed him in a difficult position. Large quantities of rice had to be imported annually on ships that the Company could put to better use. The Lords XVII were very disgruntled with this situation, and insisted that the colonists had to produce at least enough grain for their own use. This pressure from the Lords XVII at last forced the Cape authorities to take the final step. When it had become sufficiently clear that fine words could not move the colonists to till their land, they were informed by means of a proclamation in 1678 that after three months they would no longer be provided with rice.[30] In the future they had to produce their own food.

The government could not adhere to this decision, however. The grain harvest that year was nearly totally destroyed by the southeast wind and the colonists were left without any wheat for bread to support their families. Famine threatened, and the colonists once again had to be provided with rice.[31]

Admonitions and coercive measures were thus not effective in binding all the colonists to agriculture. Among the colonists as well there soon appeared a clear division into two distinct classes. One consisted of those men who continued to practice agriculture. In spite of all the hardships, they remained on the land without losing heart. Perhaps they were luckier than their fellow citizens, or produced their crops under more favorable conditions. Perhaps they were better farmers: in any community there will always be people who are more energetic, enterprising, and judicious than their fellowmen, and under similar conditions will have more success in their undertakings. To the second class belonged those who could not make a living from agriculture, or were not content with the prospects the land offered, and sought other means of livelihood.

Some were discouraged perhaps by the hard work agriculture required, and this stimulated them to apply themselves "to other, more agreeable and less strenuous occupations."[32] But there is no doubt whatever that as a result of low grain prices and high production costs "the poor again and again have tried to abandon their farming here and come to live at the fort as public house and inn keepers. . . . The harvests of the free citizens appear to decline from year to year not only on account of the above mentioned reasons," the Council of Policy explained in 1670 to Commissioner van der Broeck, but some were in such poverty they "were forced to abandon agriculture."[33]

Under these conditions agriculture at the Cape fared badly. In 1678 commander Bax reported that not even a third of the free burghers were involved in agriculture.[34] This assertion was confirmed by his successor the following year. In 1679, roughly twenty years after the two agricultural "colonies" were established at the Cape, Crudop declared that of the sixty-two burgher families of which the settlement then consisted, scarcely twenty-two could be regarded as farmers in the true sense of the word. Among the rest were a few private livestock owners and gardeners, but the great majority were town dwellers who busied themselves with activities other than farming.[35]

Some free burghers who could not make their living from agriculture applied to be accepted back into the Company's service.[36] Some free burghers concealed themselves on returning ships and fled to the motherland. Others became bakers, butchers, and fishermen, or artisans such as shoemakers, smiths, carpenters, tailors, and sawyers to try and earn their daily bread. Most, however, went to live near the fort. There they led a rather dubious existence and by their idleness, begging, and drunkenness quickly acquired a scandalous reputation.[37] Furthermore, nearly all these town dwellers kept lodging houses as a second occupation. In this way they exploited the poor sailors and other visitors, while taking advantage of a golden opportunity to carry on a smuggling trade with the ships' crews.

The government was not overjoyed at the town's growth, but in spite of all their attempts to prevent it, the number of urban dwellers continued to increase. No one was permitted to become a free burgher unless he occupied himself with agriculture, livestock breeding or "other honest permissible means of support,"[38] but it was much more profitable to run a smuggling trade or to keep a public house, than to farm. The government, which wanted to rid itself of this unproductive class, attempted unsuccessfully to persuade the town dwellers to take up farming. Finally, Simon van der Stel decided to establish a town inn, thereby hoping to compel the townspeople to take up farming by depriving them of the means to fleece visitors.[39] All attempts to get the town dwellers onto the land nevertheless failed. Agriculture was unpopular and remained unpopular.

This division of the free burghers into two socioeconomic classes for as long as the settlement was confined to the Peninsula and its environs was essentially the result of the agriculturists' unsound economic situation, and would remain a chronic problem for many years. Originally the town could offer a handful of colonists a livelihood, but it was not capable of rapid growth. Fortunately, however, there was another way out of the difficulty: stock farming, hunting for wild game, and cattle trade with the indigenous peoples offered an independent means of existence for everyone who could not or would not earn their food in any other manner.

The Initial Expansion into the Interior

The initial expansion into the South African interior was undertaken to promote agriculture. Throughout the seventeenth century it was still the farmer who led this migration up-country. This began about the 1670s. There was not much land in the vicinity of the Cape suitable for growing grain, and the government feared it could never produce enough grain there to provide for the garrison's needs. Consequently, the government decided in 1670 to send Company servants to Hottentots Holland to sow grain there.[40] This plan was implemented in 1672 and a few years later free burghers followed the Company's tracks into the interior. A few farmers, in other words, were given land on loan in the Tijgerberg region toward the end of the 1670s. With this a small beginning was made in the colonization of the interior.

Likewise in this instance, the government's intention was that the farmers should use the land for agriculture. Yet they too did not cultivate their lands properly. The government attributed this to the fact that they did not have sufficient security with regard to their right of use of the loanground; in order to encourage agriculture, it was consequently decided to grant ownership of the land to the tenant farmers involved. It was also decided to follow the same course of action if other farmers should apply for land outside the Table Valley area. But the colonists would be obliged then to cultivate the land they received in ownership. If they did not, their property rights would lapse and the land would be handed over to another person who was prepared to make better use of it.[41]

During Simon van der Stel's rule the colonization of the interior—helped along by the arrival of the Huguenots—proceeded with rapid strides. By the end of 1679 Stellenbosch was established, and seven years later there were already ninety-nine households.[42] In 1687, along the Berg River at Drakenstein, yet another new settlement was founded, with twenty-three persons receiving property.[43] In both cases the government intended the farmers to concentrate on agriculture. These two rural settlements made rapid progress. As early as 1687 the colonists in the Stellenbosch district gathered a harvest of 3,865 muids of grain, whereas the Cape district produced only 921 muids. The livestock herds of the Stellenbosch farmers were by this time also markedly larger than those of the Cape farmers. The comparative totals for 1687 were 1,605 cattle and 15,388 sheep as opposed to 1,080 cattle and 11,091 sheep.[44] Gradually Stellenbosch and its environs were recognized as the country areas, and were considered to be the actual colony. The Cape district receded altogether into the background, "and evolved in an entirely different direction, suffering not so much pain through agricultural labor as by drink."[45]

When Simon van der Stel arrived at the Cape, the thorny problem of production was not yet solved. Hence, the industrious governor did all in his power to encourage immigration and promote colonization of the interior, and thus bring agriculture to a flourishing state. Under Simon van der Stel agriculture did indeed improve quickly.[46] Land that was more fertile was brought under cultivation, the thorny labor problem was more or less satisfactorily solved, and the increased grain price spurred the colonists to greater efforts.[47] As a result the governor could report in 1688 that with God's grace he had already brought agriculture so far that there were 3,664 muids of grain in the lofts—enough for the following two years. There was also still more grain in storage at the

Company's posts and in the free burghers' houses.[48] The necessity for importing rice had thus disappeared temporarily.[49]

Unfortunately, when the Company had reached its goal the Lords XVII threatened to lower the grain price again. And in fact the price of wheat was reduced in 1690 from ten guilders to eight and a half guilders per muid. The commander was very disgruntled by this reduction. He personally thought that the colonists could not produce for a profit at under ten guilders per muid, and he feared that the fervor with which they had undertaken agriculture in the past would vanish again.[50] This did happen, and relatively quickly. The best producers could perhaps grow wheat for eight and a half guilders per muid and make a profit, and some might even have prospered, but Grevenbroeck's allegation (1695) that "all the land would be covered with gold," if each colonist performed his duties with as much attention as his rich friend,[51] is decidedly very exaggerated.

Farmers who possessed cultivated land upon which years of work had already been bestowed could naturally sell grain at a relatively low price. But it was a different matter for sons who became independent and new colonists who had to make a beginning by cultivating grain on untilled land. With great labor and expense such newcomers, who usually had very little capital, first had to clear the land. Weeds and bushes had to be cleared away, roots had to be pulled out, and it was only after one or two or three years' work that a new piece of ground—"that has possibly never known human labor or tillage since the creation of the world"—began to yield a good harvest.[52] For many beginners without property, production costs were thus higher than for the longer-established farmers. As a result they could not produce at the same price and realize a profit. This made it very difficult for newcomers to make a start in agriculture.

A new factor made its appearance in the young colony's history toward the end of the seventeenth century, and it hit agriculture particularly hard. Before Simon van der Stel reached the end of his term of office, the production problem changed to one of marketing. In 1695 the government in Batavia refused to buy the Cape's surplus grain, resulting in large stores of grain in the Company's granaries beginning to decay.[53] When the colonists saw that the Company would no longer buy any more of their grain, they proceeded to limit output to their own needs and wheat cultivation quickly fell into decline. The following year, however, notice was given that Batavia would again buy the Cape's grain, and the colonists set about the task of grain production with renewed enthusiasm. With much satisfaction the governor could report in 1697 that the colonists "have applied themselves similarly with a new desire and zeal to agriculture, plows have been put into the ground this year in far greater numbers, and also undoubtedly more wheat sown than has ever been sown before."[54]

Here we are concerned only with the beginning of the chronic marketing problem that would place restrictions on agriculture in the eighteenth century, but the above example illustrates particularly well how an uncertain market could influence grain production and cause the colonists to look for other sources of income.

Agriculture in the Country Areas

Simon van der Stel experienced the same problems as his predecessors in trying to keep the colonists on their farms. As could be expected, the greatest difficulty came from the young beginners. It appears that farmers who already had years of experience in agriculture continued with it. But again and again the government found that young colonists at Stellenbosch and Drakenstein requested land and property and received it, only to leave it lying uncultivated.[55]

This problem increased so much with the passage of time that the government was obliged to warn the colonists that their land would be taken from them and given to other citizens if they did not cultivate it within a year and a half of receiving it.[56] It is doubtful if this ordinance was ever enforced. In any case the colonists evidently did not take much notice of it. Land misuse, which the ordinance was intended to obviate, continued since new ordinances had to be issued later. And it is evident that the government was powerless to root out the evil because in later ordinances the grace period was set at three years.[57]

Other rural farmers even sold their land shortly after they received ownership of it. No legal objections could be brought against this, because originally the free burghers had received the right to lease, sell, or otherwise dispose of their land, with the reservation that this must always take place after communication with the government.[58] This privilege was abused so often, however, that a special regulation was needed in order to bind young colonists to their land. Accordingly, it was determined that except for the customary land transference tax of 2.5 percent, which had to be paid at the time of transfer of all fixed property, an additional tax of 10 percent would be imposed if the land was sold within three years, and an additional tax of 5 percent had to be paid if the transfer occurred before ten years elapsed.[59]

Some of the rural farmers who would not try their hands at agriculture, or who had abandoned it, migrated to the Cape and swelled the ranks of city dwellers. The government regarded this migration back to the city as detrimental to agriculture. In order to counter such movement, it was decided that no one would be allowed to move from Stellenbosch or Drakenstein to the Cape unless he paid fifty rix-dollars. This payment would go into the appropriate district treasury.[60] This measure, however, could at most discourage the reverse migration to the city. Like the other ordinances it could not force the farmers to cultivate their land. Nor could it prevent them from selling their land, abandoning agriculture altogether and trying something else.

Some farmers who had sold their freehold land or not tilled it nevertheless remained in the countryside. At Stellenbosch and Drakenstein there was no opportunity to keep a boarding house or to carry on a smuggling trade with passing ships. But just as Cape Town offered a livelihood for those who could not or would not live as tillers of the soil, the interior offered opportunities for others, with its good grazing lands, incredible abundance of wild game, and large livestock herds belonging to the Africans. And before the end of the eighteenth century the exodus of the hunter-migrant farmers into the interior had already begun.

The Development of Stock Farming

The first free burghers were given the right to keep as much livestock as they wished. The government even regarded private stock farming as a very important business and did its best to encourage it. For that reason, Commissioner Ryckloff van Goens gave the colonists permission to exchange livestock with the Khoikhoi provided they did not pay any more than the Company would normally have done. Shortly thereafter the commander put a stop to the exchanging of livestock, to the disappointment of the free burghers, but this measure was not directed against private stock farming. Van Goens's livestock trade concession was repealed because the colonists presented the Company with too stiff competition by pushing up the purchase price of livestock. Shortly before, however, the Council of Policy received orders from the Lords XVII to "employ all possible means to encourage cattle breeding, as this is for the time being a principal object and intention of the Company." In the light of these instructions it was then decided that in future the Company should supply the free burghers with livestock for breeding. Every free household would get up to a maximum of fifty ewes, provided that it kept at least one Dutch ram.[61]

Evidently then the colonists also had more success with their stock farming than with their grain growing. A few burghers had already begun to deliver slaughter stock to the Company by 1662.[62] By 1673 there were already burghers whose herds numbered at least six hundred animals and some were probably even greater than that.[63] According to the annual returns, in 1682 there were farmers with flocks of a thousand head,[64] and seven years later figures like 2,000 and 2,600 were recorded.[65] By the 1690s stock farming was already a very important source of wealth, and well-to-do stock farmers could be found throughout the colony. According to the 1692 returns, which were undoubtedly much too low, Henning Huising then possessed at least 5,300 sheep, while various other farmers had more than a thousand, and flocks of five hundred were quite common.[66]

The government was very pleased with the development of private stock farming. For a time the Company even tried its hand at stock farming, but not with much success. Farming on the outlying posts had to be left in the hands of slaves and servants. These men had no personal interest in the work and could not always be trusted. The Company also could not depend on the livestock trade with the Khoikhoi for their supply of meat. Since the Khoikhoi were not inclined to bring their livestock to the Cape, long and arduous trading expeditions had to be sent out after them. These did not produce enough to provide for the needs of the Company, and what is more, this trade was based on a very precarious foundation. There was evidently more or less constant warfare between the Namaquas and the Khoikhoi nearer to the Cape, in which the latter were usually defeated and the greatest part of their livestock stolen.[67] As a result, if a few sheep could be bartered now and then with the Khoikhoi, they were so thin they could hardly be used as slaughter stock.

The colonists could breed a much better grade of sheep than the Khoikhoi could supply. Their livestock was two to three times heavier than that of the Khoikhoi, and they constantly complained that they did not know what to do with their marketable slaughter animals. Under these circumstances the government decided to buy the adult wethers from the colonists at eight guilders a head. Ewes, however, had to be kept for breeding purposes.[68] In 1672 Commissioner van der Broeck suggested that no further trading expeditions be sent into Khoikhoi territory, and that all the livestock the Company needed be purchased from the free burghers.[69] The colonists could not yet supply enough sheep to meet the Company's requirements, however, and the following year it was necessary to send out another trading expedition. Still, it is clear from the above facts that the colonists devoted much attention to their stock farming, and that the government welcomed this.

The war with Gonnema made the government recognize even more clearly that private stock farming had to be encouraged. There might come a time when the exchange with the Khoikhoi would not yield anything, and then the government would be compelled to fall back on the produce of the free burghers. How much value the government placed on the colonists' stock farming is evident from the following. On 25 February 1678 Henning Huising and Claas Gerrits, two of the Company's livestock herders, requested to be released from the Company's service, and to begin stock farming in the vicinity of the Steenberg. The two livestock keepers' request was not only granted, but it was decided moreover that each of them would receive forty pounds of rice and an equal amount of hard bread from the Company during that first year.[70]

These two aspirant free burghers would farm exclusively with livestock. In that regard it is noteworthy that the government, which was doing all in its power at the time to encourage agriculture, was evidently fairly eager to grant their request, clearly demonstrating that the government regarded private stock farming equally as important as agriculture.

Stock Farming Becomes a Separate Industry

From the beginning stock farming at the Cape was more popular than agriculture. That is not difficult to understand. In the first place, stock farming was more profitable than agriculture. It cost much less to breed or exchange a sheep than to produce a muid of grain and they had about the same market value. In addition it must be noted that agricultural products could not be transported to the Cape without great difficulty and expense, while stock herds could transport themselves on their own legs to the market. If we further bear in mind that agriculture required more physical labor than livestock breeding, then it is easy to understand why many colonists would be inclined to exchange their plows for sheep. Young colonists, in particular, who had to begin without any capital, felt a very strong attraction to stock farming because it did not require a large outlay of initial capital. Young beginners had to have *livestock* with which to begin farming, but this was not a serious problem. They could easily earn livestock from other farmers, receiving a herd in return for half of the offspring, or barter for livestock from the Khoikhoi—even though the ordinances did forbid this.

As a result, stock farming at the Cape gradually gained ground, while agriculture languished. Established farmers began to give more and more attention to their stock farming and in some cases abandoned agriculture altogether.[71] Many of the young colonists in the rural areas who had received freehold property but left it uncultivated or sold it, probably also became stock farmers. In addition, many colonists gained their freedom, in one way or another obtained livestock, and began farming without having received or bought freehold land. In the beginning only their *names* appear in the tax rolls; later, however, small numbers of cattle begin to appear, increasing yearly.[72] (Compare the case of Henning Huising and Claas Gerrits.)

In this manner, stock farming, which apparently began as a sort of adjunct to agriculture, gradually became a separate industry. When this occurred precisely is very difficult to determine, because only from 1672 are the annual tax rolls compiled in such a way as to shed light on this point. But in any case, there is no doubt that about this time stock farming was already in the process of breaking away from agriculture. In the tax rolls of this year we come across the names of a number of farmers who possessed no tilled or untilled ground, who had sown nothing and harvested nothing, but who still owned livestock.[73] Thus we have here persons who at this stage already occupied themselves exclusively with stock farming. In the years that followed, the total number of men in this category steadily increased, and before the end of the seventeenth century self-supporting stock farming, as an independent enterprise, free from agriculture and the freehold, was already relatively strongly developed.[74]

Simon van der Stel and the Stock Farmers

When livestock farming at the Cape was already well on its way to becoming a separate industry, and the Cape government had already begun issuing grazing licenses to stock farmers without freehold property,[75] the colony again received a new commander. He was a remarkable man in many respects and did much for the economic development of the country, but did not have too much sympathy for the stock farmers. I refer to Simon van der Stel, who assumed his office in October 1679.

The new commander regarded stock farming as an important aspect of farming, and recognized the necessity for it, but he had constant troubles with the stock farmers. Time and again he observed among their flocks sheep without ears, in spite of the ordinances forbidding such practice, and this then involuntarily aroused in him the suspicion—in some cases perhaps well founded—that these animals belonged to the Company. In spite of his earnest attempts to protect stockbreeding, the farmers continually slaughtered and sold ewes. And when the farmers periodically separated out their rams from the flocks, the commander could not understand it, and felt that it would harm the breeding.[76]

But although Simon van der Stel recognized the importance of stockbreeding, he was above all concerned with agriculture. Under his predecessors wheat production was always inadequate, and large quantities of rice had to be imported annually from India—"yet not without the expression of great dissatisfaction"[77] on the part of the Lords XVII. Van der Stel, however, gave special attention to agriculture, with the result that, as he later proudly explained,

> this same agriculture not only has continued and expanded against all expectation on the matter, provided it is encouraged, and the farmers looked after as well, thereby not only the complaints that have arisen have been removed, but also this colony has already reached such a flourishing state, that men now by themselves can earn a sufficient living here, and also with good growth, the people of Batavia can share with us our surplus to some extent.[78]

For years Simon van der Stel strove to achieve this sound economic position, and when it finally was attained, he did his best to maintain it. Hence he considered the tendency among the colonists to limit themselves only to livestock farming as ruinous, and he opposed it with all his might. If the farmers switched over exclusively to stock farming, agriculture would decline and the good work that he had begun would be frustrated. Moreover, the commander obviously had little sympathy in principle with private livestock owners who did not know how to handle a plow and depended on others for their sustenance.[79]

Understandably, Simon van der Stel was thus very dissatisfied when it came to his notice that

> a few of these free burghers earn a livelihood by grazing livestock, having obtained the same in trade from the Hottentots, together with milk and other produce of theirs. They travel far into the interior, to the considerable detriment of the common good, and without the least bit of profit or advantage imparted to the colony, but at the cost of their leading lazy and indolent lives.[80]

In order to prevent this horrible abuse, it was decided that no one would be allowed to own livestock unless he had "a residence or land" in the Cape, Stellenbosch, or Drakenstein districts. If he did not, the livestock in question would be confiscated. Farmers who wanted to avoid such losses had to apply for plots or farmlands within eight months, had to come and live in one of the designated districts, and had to be registered at the appropriate district office. If a person did not subject himself to those regulations, then not only would his livestock be confiscated but he would be deprived forever from all provision of large and small stock from the Company. In order to facilitate the implementation of this ordinance, it was further stipulated that all livestock owners who already possessed freehold property must produce the original title deed to their land within eight months.[81]

Simon van der Stel's attitude with regard to stock farming appears very clearly from this ordinance. He wanted it to remain a secondary industry, a sort of supplement to agriculture. Nevertheless, the old governor's opposition to independent stock farming was unsuccessful. Even during *his* term of office the farmers at times had surplus grain for which there was no market. Agriculture was no longer capable of significant development, and very soon it could not offer all the colonists, whose numbers had grown rapidly, a livelihood. They could not all live in the Cape, and colonists born at the Cape had gradually begun to regard this country as their own, and thought no more of escaping to the motherland. They had to live in one way or another, however, and only stock farming was capable of expansion—right through the eighteenth century and well after that.

FROM STOCK FARMER
TO MIGRANT FARMER

Pasture for Livestock in the Interior

Finding enough pasture for the Company's and colonists' livestock became a serious problem very soon after the colony's founding. What grazing land there was in the vicinity of the Cape was sparse and very limited. During the summer months the cattle often languished in the fields and even died as a result of drought,[82] despite constant feeding with sweet hay and cauliflower leaves.[83] In 1661 the government reserved the Table Valley for the exclusive use of the Company's livestock. Otherwise it would have had to take them three to four hours' distance from the fort in order to graze.[84] In spite of this land reservation, the Company was still obliged to search for new grazing land for its livestock, and this brought with it the necessity of hiring many watchmen, which meant greater expenses. To find good pasture the livestock was eventually taken so far into the interior that herds could not be returned to the fort in the evenings. The Company was finally obliged in the 1670s to develop a system of cattle posts in the interior. Before long the free burghers followed the Company's example.

As the colonists' livestock herds increased with the passage of time and grazing areas in the inhabited parts of the colony became more limited, the farmers were forced to send their livestock deeper and deeper into the interior to find good pasture. This was especially necessary in the dry season, when there was no good pasture to be found in the vicinity of the Cape. As a result of the growing need for new grazing lands, it had also become necessary for the colonists to make use of the cattle post system, and in 1679 the government granted grazing rights to a number of farmers in the interior. Henning Huising, Nicolaas Gerrits, Jochem Marquaart, and Hendrik Elberts were permitted to graze their livestock along the Eerste River, while Pieter Visagie and Jan Mostert were given grazing rights in the vicinity of the Tijgerberg.[85]

These grazing rights were still granted during Johan Bax's term of office. When he was succeeded by Simon van der Stel a few months later, however, the issuance of grazing licenses was discontinued by the new commander. In keeping with his well-known vision of a densely populated agricultural colony, van der Stel wanted to maintain the well-cultivated freehold farm as the inexorable starting point for private farming. His position was that if someone possessed property, then he also possessed grazing rights on the land in the vicinity of his farm. If he possessed no property, then he did not even have the right to own livestock.

But the need for grazing land persisted, despite the commander's wanting to suppress the cattle post system. And when the government would not allow the farmers to look for grazing lands in the interior, they did so on their own. Thus, in time the farmers developed the practice of sending their livestock—probably most often under the supervision of older sons, servants, or perhaps reliable slaves—far up-country in search of temporary pasture.

Toward the end of the 1680s it came to the commander's attention for the first time that the colonists were letting their herds "graze on unknown and distant sites until they were fat." This was a serious transgression. He viewed it as an attempt to undermine "his ardent diligence for the well-being of this colony, to thwart him, to undercut everything, to upset his noble plans, and suddenly to smother the common good with a public calamity." It would lead to the certain ruin of the colony and, placing himself squarely in opposition to this "lawlessness," the commander strictly forbade any livestock to be allowed to graze on unknown and distant sites. There was a fine of fifty rix-dollars for the first transgression. If it occurred again, the offender's livestock would be confiscated. Corporal punishment would be meted out for a third infringement.[86]

The farmers, however, did not really take much notice of this ordinance, and the illegal activity increased rather than diminished. Simon van der Stel was aware of this, and therefore brought the matter before the Council of Policy again a few years later. It was then decided "to order and strictly command by ordinance as soon as possible" that after having grazed their livestock in the field during the day, the farmers had to return every evening to their houses. Anyone remaining away from his house with his livestock for eight consecutive days would have all his livestock confiscated. One-third of the confiscated animals would go to the Company, one-third to the fiscal, and the remaining third would go to the person that had brought in the charge. The ordinance was drawn up the same day—thus, "as quickly as possible"—but it would be three months before it was first posted.[87] Like many other ordinances, this one also remained a dead letter. The farmers would not inform on one another, and it is doubtful whether the government truly planned to enforce the ordinance, considering that it would have ruined many farmers. The civil and criminal legal rolls contain no cases where citizens were actually charged for infringement of this ordinance, or where livestock was actually confiscated.

So before the end of the seventeenth century the farmers had already begun to migrate periodically with their livestock—a practice that the climate and soil conditions of our country makes necessary up to the present day. Simon van der Stel's opposition was entirely in vain. Scarcely had he retired from his post as governor when his son and successor officially legalized migration with livestock by issuing grazing licenses. From this there later developed a fixed system of land tenure that was particularly well adapted to the needs of extensive stock farming. Despite this the arbitrary grazing on undistributed crown lands continued unabated, and the government connived at this though ordinances were frequently promulgated prohibiting it.

The Influence of the Khoikhoi on Expansion

The need for good pasture on which to graze their livestock probably would have induced the colonists to migrate inland a little earlier than they actually did, had the Khoikhoi not presented a danger in the beginning that had to be taken into account. Especially during the first few years after the colony's founding, the farmers were constantly alarmed by stock theft and the destruction of their grain fields by the Khoikhoi. In 1659 van Riebeeck's patience was stretched to the limit, and weapons were taken up to force the Khoikhoi into submission. Initially the frontier farmers simply had to abandon their fields and flee, but the following year the Khoikhoi were compelled to make peace.

Nevertheless, the Khoikhoi could not be trusted, and the colonists had to remain continually on guard against surprise attacks. This made a permanent system of defense necessary to shield the Peninsula against the Khoikhoi danger. A stone redoubt—called Coornhoop—was built in the middle of the colonists' farms to protect their grain fields against devastation by the Khoikhoi. Along the border a strong wooden fence with high posts was constructed, and in addition three guardhouses were built—called Kijkuijt, Keert-de-Koe, and Houdt-den-Bul—in order to protect the frontier. Mounted guards were stationed here to guard the colonists against Khoikhoi robberies. Even the border gates were watched over by sentries, and the "horse patrol" had to see to it that the colonists' livestock did not graze beyond the border, which was allowed only if the livestock were accompanied by a strong guard.[88]

Though van Riebeeck's vigorous policy filled the Cape's native inhabitants with respect for the Dutch, these defensive works were not excessive because at the beginning of the 1670s another war broke out with Gonnema that lasted about three years. His people attacked the free burghers, burned their houses, and even murdered a woman. The Company's post at Saldanha Bay was plundered as well in 1673, and the post keeper, a soldier, and two free burghers were murdered. In the preceding year the Khoikhoi took the lives of two hippopotamus hunters and burned their wagons. In the following year again the Khoikhoi cornered eight big-game hunters.[89] Thus, by the early 1670s the Khoikhoi had made the interior unsafe even for hunters, and it is therefore easy to understand why an isolated stock farmer would not take lightly the risks associated with venturing into the wilderness.

The war against Gonnema was the last declared conflict between the colonists and the Khoikhoi but the free burghers' fear of the Khoikhoi disappeared only a number of years later. When grazing rights along the Eerste River were granted to Henning Huising in 1679, he was warned not to quarrel with the Khoikhoi, who also used the pasture in that region.[90] Fear of the Khoikhoi was also one of the reasons why Simon van der Stel allowed the farmers in the Stellenbosch area to live more or less together. Even in 1687 when he parceled out land along the Berg River, he first found it necessary to confer with the Khoikhoi captains. It was sometimes necessary even later to let fertile ground lie fallow in order not to weaken the lines of defense too much.[91]

The Khoikhoi did not long remain a serious counterforce against the farmers' migration into the interior, however. While the white population grew and their strength increased, the might of the Khoikhoi gradually declined. Their internecine wars, as well as attacks from the Bushmen, contributed much to this decrease, but undoubtedly it was the white man's diseases that he brought with him to the Cape, and against which the Khoikhoi had no immunity, that also helped thin out their numbers.[92] Theal tells, for example, of a sickness that struck the Cape in 1687. Many whites lost their lives, but the devastation that it caused among the Khoikhoi was much worse. "They suffered severely from it, so much that one kraal is mentioned in which half the people were dead while others were all sick."[93]

This sickness perhaps broke the power of the Khoikhoi. In

any case, long before the first smallpox epidemic in 1713 they were
no longer a threat to the settlement, and had left open the path
into the interior for the migrant farmer. In 1688 Simon van der
Stel reported that there was no longer any need to be fearful of
the Khoikhoi. On the contrary, the colonists could expect nothing
but loyalty and good service from them. The captains asked the
governor, who evidently dealt with them very tactfully,[94] for his
friendship and protection: "their love and affection toward us growing more and more, so that at present during the pressure of harvest
and plowing time, they slip down among us just like the West-
phalians in the Netherlands."[95]

The Livestock Trade with the Natives

Originally the free burghers were not allowed to trade for the Khoikhoi cattle, but Commissioner Ryckloff van Goens very soon gave
them permission to engage in bartering. Their only obligation was
not to pay any more for the cattle than the Company normally did,
and to buy from the Company, and no one else, the trade goods
they needed for the exchange.[96]

The free burghers proceeded to take advantage of this excellent
means of enlarging their herds. It became clear very quickly that
they were more interested in the barter than in their farm lands.
When the government organized a trading expedition in October
1657 to the Saldanhers, the free burghers asked to go along. Their
request was granted. The expedition was led by Abraham Gabe-
mma and there were eight free burghers, seven Company servants,
and four Khoikhoi who took part. The Company provided the
trade goods; the colonists had to furnish their own food and ammunition and they were to receive one-third of the bartered stock.[97]

Scarcely had this trading expedition returned before the colonists wanted to go off on another one—this time on their own.
Their request was granted and ten days later they were back. The
trip had not produced much. They brought back a thin cow, three
calves, and forty-seven old sheep, and the experiences they had
were "such that the free men were given a fright, and have no
intention of going out again on their own."[98] Their fright must
have passed quickly, however, or given way to the desire for adven-

ture and profit, because by the month's end van Riebeeck had to exhort them earnestly not to "go out" again. The commander suggested rather that they remain on their land to harvest their wheat and ensure that it was brought home, "and that instead of all that roaming about, to take care that their houses and barns were made ready for the storage of the said grains."[99]

The following year the stock trading concession was revoked because the free burghers offered too keen competition for the Company, which was also interested in the trade. But the ordinances prohibiting these exchanges could not constrain the desire for trade that stirred in the colonists' blood. They had always belonged to a nation that thought of themselves as the best traders in the entire world and their unsurpassed desire for trade was glorified in the proverb: "For profits the Dutch merchant would sail through hell, at the risk of scorching his sails."[100] And stock trading with the Khoikhoi was not so dangerous, in spite of the ordinances. Thus the colonists carried on with their trading activities.[101]

A number of precautionary measures were taken to prevent a smuggling trade with the Khoikhoi. The colonists were obliged to give a regular account of the lambs that were born, as well as of the sheep they had slaughtered. The Khoikhoi were forbidden to visit the colonists' homes,[102] and neither the colonists nor their servants were allowed to visit the Khoikhoi kraals. They were even forbidden to speak to the Khoikhoi. No one was permitted to go hunting outside the area patrolled by the mounted guards. The colonists were also forbidden to allow their livestock to graze out of the sight of the soldiers in the watchtowers.[103] If Khoikhoi appeared, offering to barter their stock, the free burghers were to direct them to the fort. The Company would then trade for the stock, and the colonists could then buy it back again at the purchase price. This would keep the prices low! Furthermore, the punishments for transgressing the trading ban gradually became more severe, and after 1677 corporal punishment was even meted out.[104]

All these ordinances, however, were of no help. The colonists went to trade entire flocks of sheep at the Khoikhoi kraals, or sent Khoikhoi who were in their service to do so.[105] At night they slipped by the mounted guard and undertook adventurous excursions to distant Khoikhoi kraals. Or, by day, they would drive their stock a little farther than normal into the countryside, meet the

Khoikhoi at a previously agreed on spot, mix the bartered stock with their own and return home in the evening as if nothing had happened.[106] And if they found it impossible to bring the stock home alive, they had them slaughtered at the Khoikhoi kraals and smuggled the meat past the guardhouse,[107] despite the fact that the mounted guards stationed there were to search all wagons that passed by.

The Khoikhoi welcomed the colonists as traders, because they generally paid a much better price than the Company did. In 1677 the government claimed that the colonists were paying eight to ten times more than what the livestock was worth.[108] Consequently the Khoikhoi were careful to do whatever was necessary in order to sustain this trade with the colonists. At the same time, the Company's trade was steadily falling off. When Simon van der Stel resigned his office, affairs were already so serious that the captain of the Hessequas refused point-blank to trade with the Company, and the other tribes frankly acknowledged, "that they were much more favorably disposed toward the burghers" than toward the Company.[109]

When the neighboring Khoikhoi no longer had enough livestock to exchange, the colonists of Stellenbosch and Drakenstein began making expeditions deep into the territory beyond the mountains. They migrated out together in groups and carried with them large quantities of tobacco, copper, beads, and other trading items. Sometimes they even pretended to be Company servants and forced the Khoikhoi who were unwilling to trade to do so, in the name of the governor.[110]

The government soon realized it actually had no control over the traders once they had crossed the mountains. It consequently forbade the colonists to "proceed over the mountains of Hottentots Holland, over the Rode Sand, Elephantspad [present-day Franschhoek Pass], and in general over any chain of mountains to the Sousequa, Hessequa, Ubiqua, Grigriqua, or Namaqua Hottentots whether alone or in the company of others, with or without wagon or livestock."[111] Any person found beyond these mountains would be fined sixty rix-dollars, be sentenced to a year's hard labor, and, in addition, have all of the goods and livestock found in his possession confiscated.[112] To enforce this ordinance, however, was practically impossible because the government could not keep watch over all the mountain passes.

The punishments imposed in the old days for transgression of the laws were often horribly cruel. For example, in 1692 two Stellenbosch burghers were summoned before the Council of Justice on charges that they had shot big game, bartered with the Khoikhoi for sheep, and stolen some of the Company's livestock. Both were found guilty. The first was condemned "to be whipped with a sheepskin hanging over his head, a fine imposed of a hundred rix-dollars—half for the Council Chamber, and the other half for the plaintiff, and banned for the time of six years on Robben Island at public labor." The other's punishment was "also a fine of a hundred rix-dollars to be divided as before, and likewise banned for a period of three years on the same island at public labor. In addition, the condemned are to pay court costs and fees."[113] These penalties naturally applied to a variety of transgressions, the worst of which was the theft of Company cattle. In 1689, for example, five Stellenbosch citizens were found guilty merely of contravening the cattle trade prohibition, and each was fined only twenty-five rix-dollars, together with confiscation of the wagons and cattle on behalf of the Company.[114] Nevertheless, these sentences were severe for the crime for which they were imposed.

The arm of justice did not reach very far, however, no matter how severe and relentless the law was. Undoubtedly the number of persons summoned before the Council of Justice for violating the trade ban was extremely small in comparison with the number of farmers that took part in the illegal barter. It was nearly impossible for the magistrate to catch the stock traders. Occasionally men were summoned before the Council of Justice, but in most cases they were acquitted because it was impossible to find evidence of their guilt.[115] Consequently, there were seldom any cases where traders were actually punished. And even though these punishments were often severe, they did not deter others from the livestock trade because the chances of being caught and receiving punishment were so slight and the prospects for profits were so great.

The Influence of the Livestock Trade

The livestock trade with the Khoikhoi had a particularly important influence on the development of the migrant farmer. The colonists had not only become acquainted with the interior on their trade expeditions inland, but the profits that were attached to the trade and the adventure that went with it had cast a spell on them. It had already diverted the established agriculturists' attention away from their work and in the direction of stock farming exclusively. Although by the end of the eighteenth century the farmers were all still settled in an area that was more suitable for agriculture than for stock farming, and where production for the Cape Town market was still economically feasible, many of them had already begun to sow just enough as was necessary for their own needs, and for the rest of the time were applying themselves to stock farming. And while the magistrate and militia officers, appointed in 1693 to bring charges against the lawbreakers,[116] perhaps could have caused trouble if livestock was bartered, such farmers had settled as far as possible outside the inhabited parts, "so that under that pretext, they sought their additional subsistence by exchanging livestock, butter and milk with the Hottentots, and that became their principal occupation."[117] Other farmers went even further. They had not only exhausted their land and sought the "greater part of their subsistence," through forbidden trade, but had even "completely abandoned their obligatory cultivation, together with their depleted farmlands."[118]

If the livestock trade caused so many of the already established farmers to divert their attention away from agriculture, then it is easy to understand why it had an enormous influence on the activities of those colonists who still had no established interests. While it was terribly difficult for a beginner without capital to take up agriculture, the livestock trade afforded him the opportunity to build up a strong stock farming enterprise within a relatively short time. It was evidently for this reason that so many young colonists received freehold property but never worked it and later even sold it. In addition, some people often became free burghers under the pretext that they would practice agriculture, but then they never settled permanently anywhere. They simply wandered about, living now here and then there with one or another free burgher and in the meantime going out to exchange livestock for their patron.[119] It is evidently these persons whose names often appeared on the tax rolls after 1672—persons who had no possessions in the beginning later declared small herds of livestock and then became wealthy stock farmers without ever possessing private property.[120]

Finally, the livestock trade still offered a livelihood for Company soldiers who had deserted and fled into the interior, often to avoid a sentence of death. They usually sought refuge in the mountains, bartered for livestock with the Khoikhoi, and then sold them again at a low price to the colonists.[121] So long as they remained in the interior they were comparatively safe. The farmers seldom informed against them. Such runaway soldiers also evidently occupied themselves later with stock farming.

In this way the livestock trade with the Khoikhoi advanced the development of stock farming and, before the end of the seventeenth century, contributed thus to the appearance in the interior of a half-nomadic, pastoral population.[122]

Game Hunting

Just as there were independent fishermen and woodcutters among the first colonists, there were also a few independent hunters who had the exclusive right to hunt and to sell meat. These were never large numbers of men, however, and they had little influence on the development of the migrant farmer. Of much greater importance was the fact that all colonists were hunters to a certain extent. In those days South Africa was a true hunters' paradise and all the colonists were good shots. They had carried their weapons in the motherland, as well as in the Cape, and knew "about the wonders that accompanied it." In their constant struggle against lions and leopards and other savage beasts of prey, they became more skillful every day. To his surprise, Simon van der Stel noticed at the militia practice at Stellenbosch that while the militia were shooting at the target and the wooden parrot, "of a hundred shots scarcely one missed its mark."[123]

Under these circumstances hunting for game soon became extremely popular. Already by the beginning of the 1670s, or perhaps even earlier, the colonists had developed the habit of traveling up-country in their wagons to shoot big game to support their families and slaves.[124] During the war with Gonnema the farmers were afraid to go hunting for hippopotami, but in 1677 the commander made it known that it was safe to go shooting again since a lasting peace had been concluded with the Khoikhoi.[125] After that the hunting again continued as before. By about 1686 the hippopotami in the environs of the Cape were already so decimated and driven away that the colonists had to go on eight, ten, or even twelve days journeys—even to the banks of the Olifants River—before they could get their hands on the much sought after fat.[126]

The inevitable result of this constant hunting was that the game was quickly being depleted. The government even began to fear that the game would be totally annihilated. Consequently, hunting was forbidden during specified times of the year, and even when it was permitted the colonists were obliged to take out hunting licenses. In practice, however, all the hunting laws were ineffective. The government never had enough administrative power to enforce the laws, and farmers shot game whenever they had the chance. When the hunting season was closed they even slipped pass the watch at the Company's outposts at night to go hunting for hippopotami.[127]

By this time Simon van der Stel already regretted the fact that the proclamations, statutes, and ordinances "promulgated from time to time for the welfare and promotion of this colony and its inhabitants, and posted in the usual places, in the course of a few years were either entirely forgotten, or fell into disuse for other reasons."[128] His successors had the same experience. In particular, specific types of legal transgressions, which did not conflict with the popular ethos, occurred quite often among the pioneer population. Although the law forbade it, they hunted for game, carried on a smuggling trade with the Khoikhoi, and migrated over the borders with their livestock in search of grazing land. And they did it without any feelings of guilt that they had committed a crime—just like many otherwise law-abiding citizens today will shoot game when the hunting season is closed, or take part in the illegal diamond trade if they think they can do it without getting caught.

The colonists maintained that they shot game to provide for their families and slaves. This was evidently not too far from the truth, for right through the course of our history the pioneers on the borders more or less lived by hunting and saved their livestock for the market. But often the hunt was also closely bound up with the pleasure that it offered. Undoubtedly the hunt for game was practiced as an adventurous sort of sport by many. Simon van der

Stel, who had even less sympathy for the hunters than for the stock farmers, rightly observed with reference to hippopotamus hunting that "their meat would be much more expensive if the marksmen remained at home and kept to their work; moreover, it is certain that the shooting of these beasts was pursued more for recreation than for profit by the citizens."[129]

The detrimental consequences of this continual hunting also became quickly evident. The game were gradually decimated and driven away, "while many of the inhabitants take time off from their daily work to play the sluggard and simply abandon their agriculture."[130] Agriculture did not offer many prospects, and certainly not all the free burghers were cut out for it. In addition, the nomadic life they had led in the Company's service did not help them to adapt to the daily routine of life on a farm. It need hardly come as any surprise, therefore, that the roving life of a hunter had to have had an irresistible attraction for those free burghers seeking a more adventurous life. Besides, the hunt for game naturally produced a certain income as well, and in addition it placed the colonists in the position of being able to exchange livestock with the natives. Undoubtedly there was a close connection between hunting, the illegal livestock trade and the development of stock farming.

It is very difficult to locate precisely the hunting areas that were visited by the first pioneers. During the early years nearly all the hunting licenses that were registered in the Old Gamehunters Books failed to specify a particular hunting area. In addition, not everyone who went hunting—particularly the expert hunter—took out a hunting license, and undoubtedly those who went hunting without permission had the greatest influence on the development of the migrant farmer. It was almost always from precisely this more or less adventurous element of the population, who stood relatively free from established society and remained unknown to the government, that the first migrant farmers came. It appears from the Old Gamehunters Books, however, that the first colonists went hunting nearby the Cape, especially in the dunes behind Blaauwberg. When the game became more scarce, they had to travel further inland. The hippopotamus hunters followed the rivers, and by the 1680s they had gone already as far as the Olifants River. The majority of farmers, however, had not penetrated so far inland. Toward the end of the seventeenth century it seems the most popular hunting areas lay in the direction of Groene Kloof, Riebeek-Kasteel, and the Twenty-four Rivers. About this time there was a considerable rise in the number of licenses for the Little Berg River and Roodezand areas.[131]

Apparently these hunting areas were visited only by hunters initially. But before long, in the wagon tracks of the hunters, who were the trailblazers for white expansion into the interior, came the migrant farmers with their livestock. This connection between hunting and the stock farmers' migration inland appears obvious from the fact that the first grazing licenses in the eighteenth century were issued in the very same areas that were visited the most by hunters toward the end of the previous century. It is perhaps also noteworthy that grazing and hunting licenses at the beginning of the eighteenth century were registered together in the Old Game-hunters Books, and that those books retained their original name, even after they came to be used only for the recording of grazing licenses.

In the old days, apparently, the stock farmer and the hunter were one and the same person. It was therefore quite natural that the stock farmers should migrate to the choice grazing lands that they had discovered on their hunting expeditions. And those grazing lands were naturally doubly attractive because in the interior there was an abundance of game. Thus, in the early years hunting for game not only diverted the agriculturists' attention from their work, but right through the course of our pioneer history the game beyond the borders also enticed the stock farmer to migrate inland.

Summary

Thus, in a relatively short time after the settlement's founding, different factors contributed to a large number of free Cape colonists seeking, and finding, a livelihood outside agriculture.

Under the commercial monopoly system the Company was the only purchaser of the colonists' produce. Prices were not determined in a natural manner through supply and demand, as in a free market, but were fixed by the buyer. And unfortunately for the colonists, the Company was a commercial company that was interested in immediate profits, not colonization. As a result, prices for agricultural products were kept so low that agriculture was not profitable. In these circumstances the colonists began to seek other ways of making a living outside of agriculture, even before the chronic marketing problem that began toward the end of the seventeenth century as a result of overproduction limited agricultural development still further. Moreover, as the colonists established themselves ever deeper in the interior, the transport of agricultural products became more difficult and also brought with it additional expenses.

Whereas even for established farmers agriculture lacked every element of economic attractiveness, stock farming offered better prospects. Meat could always be sold, production costs were substantially lower in stock farming, and what is more, a man could begin stock farming immediately and almost without any capital. This latter factor was of great importance for the young beginner, since sons who became self-supporting and Company servants who were discharged usually had to start with nothing.

Apart from these factors, over which the colonists had no control, there were also certain diversions that drew the attention of established agriculturists from their work, and so absorbed the time of many young colonists that they never even considered giving agriculture a try. In the beginning the temptation to go and live nearby the fort, to keep a boarding house, and to carry on a smuggling trade with the passing sailors, was too strong for many free burghers to resist. In the country areas the hunt for game, which played an important role both as an economic activity and as a sort of adventurous sport, as well as the livestock trade with the Khoikhoi, which was undertaken with an eye on profits, had the same results. It was just as difficult for the hunter or the livestock trader to be a good agriculturist as it was for the boarding house owner, but the former could continue stock farming and did so more often than not. This was especially true since hunting generally went hand in hand with the livestock trade, and the trade was such an easy means to build up one's herd.

All these factors resulted in the settled farmers occupying themselves more and more with stock farming, until they were neglecting their agriculture altogether, while many young colonists began directly with stock farming, without first trying their hand at agriculture. In this way independent stock farming, which stood free from the ties of agriculture and freehold property, was already developing by the 1670s.

As a result of the expansion of stock farming there soon developed a shortage of grazing land in the inhabited parts of the colony. Moreover, grazing land in the summer—the dry time of the year— was always of poor quality in the vicinity of the Cape, Stellenbosch and Drakenstein, and there were also occasional droughts. As a result of all these factors farmers were forced to go inland in search of new pasture and the migration with livestock soon became a necessity. In the beginning the farmers evidently migrated only periodically with their livestock, but it was not long before the livestock herds remained behind for good in the new grazing lands. Stock farmers who possessed no freehold property in the inhabited regions chose to live in the interior because there were vast and good pastures—perhaps even in order to be outside the reach of the government, which confiscated their livestock if they were not "housed and propertied." And apparently even farmers who were settled in inhabited areas in the seventeenth century already had livestock posts in the interior, where their farms were under the supervision of older sons, servants, or perhaps even trusted slaves.

Undoubtedly the frontier migration was caused primarily by the scarcity of good grazing land in the inhabited regions of the colony, but the possibility of hunting game and trading livestock with the Khoikhoi certainly contributed to entice the stock farmers into the wilderness and to keep many of them there. The hunt often provided meat for daily use, and naturally a little money could be made also from the sale of skins and horns. The forbidden livestock trade not only produced high profits, but for beginners it was the most effective method of increasing their herds. And fortuitously there was not only good pasture in the interior but also game and Khoikhoi who owned livestock.

Thus the combination and interaction of different factors on the borders of the agricultural colony at the Cape before the end of the seventeenth century brought about the development of the half-nomadic pioneer type, who led the quiet expansion into the interior in the following century. The pioneers who traveled beyond the borders can be divided for the sake of convenience into three classes—hunters, livestock traders, and stock farmers. But that division cannot be too strictly maintained. Only in a few cases would one have found the different types in a pure form. There were all sorts of combinations and intermediate forms to be found, and undoubtedly the hunter, the livestock trader, and the stock farmer were more often than not the same person.

So far as their origins are concerned, these pioneers also belonged to different classes. Some of them were evidently farmers who abandoned agriculture and went in search of a livelihood in the interior. Others were certainly sons of settled farmers or servants who began to farm independently. Some were persons who obtained their freedom under the pretext that they would practice agriculture, but who sold their land or just left it unworked—or had not received land—and then began stock farming. In some cases therefore the agricultural farmer became a stock farmer, who in turn became a migrant farmer, who was at the same time a hunter and livestock trader. But naturally the opposite could also have occurred. Certainly all the pioneers did not pass through the same evolution.

THE FIRST MIGRANT FARMERS

Ordinances against the Inclination to Migrate

One of the things Simon van der Stel found most distressing was the inclination to migrate that had become so prominent among the stock farmers toward the end of the seventeenth century. It was particularly disturbing to him because he was opposed in principle to the idea of a rapid expansion into the interior.

To a great extent, political motives formed the basis of his idea of allowing the Cape colony to develop into a densely inhabited agricultural settlement. He would gladly have seen the total of able-bodied free colonists increase to two thousand. Then they would be capable "of defending against all landings, hostile attacks, and suchlike, from European rulers, in such a way that the people here should have no fear of enemy assault, or attack from some European sovereign or potentate."[132]

The dispersal of farmers into the interior would upset this plan. They had to live nearby each other so they could be easily warned and gathered together in time of danger. The commander even deplored the fact that the colonists at Tijgerberg and Stellenbosch were living a little too distant from one another. Consequently, he took special care to see that the farms distributed around Drakenstein were nicely placed alongside each other. Furthermore, through a systematic distribution of land he intended to unite the farms all along the Berg River up to Saldanha Bay into a single line of communication.[133] In this way the colony would eventually be raised to the desired state of defense.

The old governor was also opposed to the colony's rapid expansion because that would have led to the neglect of agriculture, the thing most dear to his heart. As a result he handed over the colony in 1699 to his son and successor with much apprehension, and gave him fatherly advice on how to act toward the colonists in general and the stock farmers in particular. Viticulture had to be controlled in some degree, as well as the consumption of alcohol. Both led to the neglect of agriculture. In general, the farmers did not display enough industriousness in their work. When the old farmlands became exhausted they wanted new lands, under the pretext that they did not have enough cattle to provide manure for their fields. But Willem Adriaan must not take any notice of these tricks, otherwise the farmers would very quickly find South Africa too small for their needs![134]

If the agriculturists had already aroused the governor's concern, then it is easy to understand why he would consider the presence of stock farmers beyond the borders as pernicious. Not only had these men "done considerable harm to the common good, without bringing any benefit and advantage to themselves," by leading a life of ease, but they were also no longer under the control of the government, and that could give rise to a smuggling trade with the natives. Moreover, Simon van der Stel was not altogether trustful of the Khoikhoi. If the colonists now were scattered far and wide from one another, the Khoikhoi could attack them one after the other and rob them of "life and property," without their being able to call each other for help and before the government could get wind of it.[135]

In order to prevent all these irregularities, the government then issued an ordinance advising everyone that all free peoples outside the boundary posts, or borders of the Cape territory, and that of Stellenbosch, together with those settled at Drakenstein, or settled round about there with their livestock, should break up their camps as quickly as possible within the next six months, and by this date have settled themselves within the proper limits with goods and chattels, on pain of corporal punishment as deserters and vagrants, and their houses, herds, and cattle pens subject to confiscation at their own expense.[136]

The Migrant Farmer at the Beginning of the Eighteenth Century

It is not known if Simon van der Stel's migrant farmers—the half-nomadic hunter-stock farmers that by the 1690s were roaming about beyond the boundaries of the settlement and exchanging livestock with the Khoikhoi—responded to the above-mentioned ordinance and returned to the colony. It is, however, doubtful. And even if it happened, then shortly thereafter other migrant farmers replaced them, because by the beginning of the eighteenth century there were already a substantial number of migrant farmers in the interior and their number increased rapidly after 1700.

This migration of stock farmers inland was also no longer viewed negatively by the government. This was due to Willem Adriaan van der Stel's having succeeded his father as governor in 1699. At this point a new era dawned in the economic history of our country. In contrast to his father, who could see little value in extensive stock farming, the new governor was preeminently the great stock-farming pioneer. He did not oppose the stock farmers' inherent inclination to disperse themselves over a large area, and was even an advocate of expansion into the interior. From the beginning of the eighteenth century the stock farmer, penetrating inland behind his large herds, came more and more to the fore. And though the government was not exactly sympathetic to those pioneers, at least it no longer interfered with them.

Shortly after his arrival at the Cape, Willem Adriaan van der Stel went on a month-long inspection tour throughout the colony in order to inspect the Company's outposts and to examine the condition of the colony in general. He found, among other things, that "some inhabitants from Stellenbosch and Drakenstein possessed too little land to maintain their livestock, and also had insufficient space to have a wide choice of land on which to continue stock breeding and wheat farming."[137] Drakenstein was a "bad and watery land," he wrote on another occasion to the Lords XVII, "but for the most part these people (the inhabitants) over there live too near the others, and consequently the majority of them could not move away very easily, yet many are on the point of being reduced to poverty."[138] Apart from these, there were in the colony various persons that were still not housed—among others, a number of people who had just recently arrived from Holland[139] and who were in need of land in order to practice agriculture and stock breeding.

It was in these circumstances that the governor's eye fell on a "beautiful land lying about eighteen to twenty hours from this Castle." It was the present-day Tulbagh Valley, situated between the Obiqua and the Witzenberg. Not only the beauty of the setting but also the "quality of the ground, both for farming and for the grazing of livestock," made a particularly favorable impression on the governor, and in spite of the fatherly advice of his predecessor he decided to give this strip of land to the propertyless colonists, "and in the course of time to extend the boundaries of the colony in this way."[140]

Willem Adriaan van der Stel's break with the economic policies that his father had followed for twenty years can be ascribed to various factors. When he became governor the thorny problem of production, which had so tormented his father, was not only solved, but it had in fact become more difficult each year to find an outlet for agricultural products on the limited Cape market, which was not really capable of expansion. A chronic marketing problem now threatened the colony. It allowed Willem Adriaan to grasp the important role that stock farming would play in the future. Not only was there always a greater demand for meat than for grain, but stock farmers were also less dependent on the market.

Furthermore, with the passage of time yet another of Simon van der Stel's objections to stock farming, and the expansion into the interior that accompanied it, had become anachronistic: the Khoikhoi no longer constituted a political threat to the young colony. In the Land of Waveren there were very few, if any, Khoikhoi to be found. Deaths had appreciably reduced their number. Those that remained were totally impoverished and had moved deeper into the country. Therefore, the governor could see no harm that might come to the colony if it were to grow. Accordingly, the first farmers departed for the Land of Waveren toward the middle of 1700 to clear the land, while the wives and children temporarily remained behind.[141]

It is worth noting that this expansion into the Land of Waveren still took place on the basis of the freehold farm, intended for agriculture. It was, however, for the last time. Hereafter the stock farmer would proceed ahead in order to occupy the land on the basis of grazing licenses, which primarily satisfied the needs of stock farming. These stock farmers frequently sowed grain on their livestock posts for their own use, but they remained primarily stock farmers. The agriculturists followed later. Even in the Land of Waveren, where permanent colonization occurred on the basis of the freehold farm, the colonists applied themselves from the outset almost exclusively to stock breeding, and produced "little wheat and hardly any wine."[142]

In theory ordinances forbidding wandering about in the interior with livestock were still in full force. Willem Adriaan van der Stel had not formally repealed them. On the contrary. All the provisions appeared again in the general ordinance of 1705: those persons who had no "house or property" in the Cape district or the districts of Stellenbosch and Drakenstein could own no livestock. The colonists of Stellenbosch and Drakenstein had to return to their homes each evening with their livestock after they had grazed their herds in the fields by day. Those who had settled beyond the boundaries or were situated there with their livestock, must return within six months.[143] These provisions prove nothing regarding Willem Adriaan van der Stel's attitude toward the migratory movements of the stock farmers, however. It was common practice every year to issue a codification of all the most important laws in the form of a general ordinance, and this was apparently done by a subordinate official who did not always ask himself whether the government still intended to keep all these laws in place. By his system of grazing licenses Willem Adriaan provided sufficient evidence that he did not view migration with livestock as an evil. The qualification for livestock ownership was later changed as well. Instead of having to possess freehold land, a livestock owner had only to be registered as a citizen.[144]

After the repeal of the ban on livestock trading in 1700 the colonists' livestock herds increased rapidly. Already in Kolbe's time wealthy stock farmers could be found everywhere in the colony. "It must be a poor settler," reported that writer, "who does not possess at least six hundred sheep and one hundred head of large cattle. When he has his affairs even a little in order, however, one quite frequently finds a thousand and even some thousands of sheep and two hundred to three hundred head of large cattle owned by one man alone."[145] Valentyn gives yet an even more rosy picture of the stock farmers' prosperity at the beginning of the eighteenth century. He assures us that he had known farmers who possessed ten to twenty thousand and more sheep.[146] Valentyn exaggerated or generalized a little perhaps on the basis of a few examples, but if we examine the yearly tax rolls it is evident that by 1700 stock farming at the Cape was already a robustly developed industry and that during the first decade of the eighteenth century it expanded surprisingly quickly. Compare for example the following table, drawn up on the basis of information from outgoing letters:

Date	1700	1705	1710
Adult males	458	568	656
Cattle	8,357	11,964	20,082
Sheep	53,971	76,423	131,630

According to this table, every adult male in the colony in 1710 must have possessed on average about thirty cattle and two hundred sheep. Many of these men were, however, town dwellers and craftsmen who possessed no livestock, while others were farmers with a minimal amount of livestock. In addition, if we make allowance for the fact that the colonists commonly gave much lower estimates of their possessions in order to avoid higher taxes, then it is quite possible that Kolbe gave a more or less accurate picture of the stock farmers' wealth.

The growth of the colonists' livestock herds inevitably resulted in grazing in inhabited parts becoming much more restricted. By 1705, during the rainy season, there was scarcely enough pasture and water for the livestock. In the dry spells there was sometimes no pasture and water to be had at all. And this state of affairs had grown so much worse from year to year that the colonists "already had to go very far with their livestock to graze."[147] To return to their homes in the evening with their livestock, particularly in the dry season, was impossible. It had made seasonal migrations on a more or less regular basis increasingly necessary—something which, according to Bogaerts,[148] was already a common practice by the beginning of the eighteenth century, the extent of which is undoubtedly very poorly reflected by the number of temporary grazing licenses that were issued during the first years of the eighteenth century.

As a result of the rapid development of stock farming it now became necessary for the colonists to live farther apart from each other as well. In the future, wrote Willem Adriaan van der Stel to the Lords XVII, the farmers would be obliged to "live very spread out, and to take their retreat at a considerably greater distance." Neither the migration with livestock nor the scattering of the stock farmers over a great area troubled the governor, however. It was not only inevitable, but was also in the Company's best interests. He wrote to the Lords XVII,

> It was truly experienced and fully understandable, that since the free population with their livestock have migrated so far inland, where there is much better grazing than there is around this region, the sheep and cattle are increasing in number, and becoming fat and healthy, such that the meat nowadays is better than before, and can be bought much more cheaply as well. If we were now to order these inhabitants to return, they would not know what to do with their livestock, because the land, as at Stellenbosch, Drakenstein, as well as here around the Cape, is mostly all distributed, occupied, and laid bare from overgrazing. From this then it also has to follow that the livestock, instead of multiplying and becoming fatter, shall become emaciated and die away, to the great detriment not only of the inhabitants, but also to the inconvenience of the Company, for meat will certainly then become more expensive.[149]

The Living Conditions of the First Migrant Farmers

There is, unfortunately, very little known about the living conditions of the stock farming pioneers who first began the migration into the interior. By the early 1690s a few extensive stock farmers were roaming about with their livestock beyond the settlement's borders, and according to the ordinance of 19 October 1692, one could conclude that some of them were already established there and had set up "houses as well as sheep and cattle pens." Moreover, it seems that these farmers were making a living exclusively from stock farming, and for the rest they bartered livestock, milk, and other goods from the Khoikhoi. We do not know any more, however, of these earliest migrant farmers. The first report describing the living conditions of the pioneers on the colony's borders dates from the beginning of the eighteenth century.

By this time already one could come across hunter-stock farmers in various places in the interior. For example, at Riebeek-Kasteel, Twenty-four Rivers, the Heuningberg,[150] as well as at Groene Kloof and Roodezand, where the farmers "cultivate no grain or wine; they do have, on the other hand, very fine pastures for cattle and afford the best summer grazing lands since there is enough water and grass."[151] These pioneers lived altogether isolated from the outside world, and all sorts of geographical obstacles impeded communication with Cape Town. Reaching Roodezand, for example, was very difficult for it was practically impossible to cross the steep Obiqua Mountains with a wagon.[152] Kolbe, who sojourned at the Cape from 1705 to 1713, related that the first pioneers who migrated there had to unload their wagons at the foot of the mountains, take them apart, and then carry the freight as well as the wagon piece by piece to the top![153]

Under such circumstances it is easy to understand why the pioneers could not bring with them many of the little comforts of civilized life. Nearly cut off by distance and transportation difficulties from Cape Town—the market and business center at the time—the pioneers had to maintain a primitive lifestyle, which was oriented toward providing for the basic necessities of life. They knew neither wine nor beer. Water was the daily and general drink. Even milk was a luxury, for which they traded with the Khoikhoi in order to have some occasionally on Sundays. Honey, however, was abundant. The colonists also traded for honey with the Khoikhoi, who scrambled up the steepest precipices to get hold of it, and they were good at knowing how to deal with the bees. The honey was then used by the pioneers to make a sort of beer, which constituted a healthy and, if desired, potent alcoholic beverage.

Most of the time the pioneers went without bread. They regarded it as too much trouble to sow wheat themselves. To transport flour over long distances and rugged trails was too difficult; and of course, they also did not have money to buy it. "So they eat flesh between flesh, that is, they take a piece of mutton or lamb and eat, instead of bread, dried hartebeest or other game meat for that purpose, and this proves such a healthy diet for them that one very seldom hears of any illnesses.[154]

Every day the pioneers went out hunting for wild game because it provided adventurous relaxation and food. Even wealthy stock farmers frequently lived more or less from hunting, because by doing so they could save and breed their livestock. And from the offspring they sold the wethers, which enabled them to purchase gunpowder and other essential necessities of life. Meanwhile, the cattle provided dung, which was used in place of firewood wherever the latter article was hard to find.

The combination of game hunting and the requirements of extensive stock farming kept the pioneers in a state of great mobility. The fact that game was continually roaming about, together with the geographical configuration of the land, made it desirable—often necessary—for the stock farmers to change their dwelling places. Thus, in the beginning, when the world still lay open and unoccupied, the pioneers did not maintain fixed abodes. Just like the nomadic Khoikhoi, they wandered about from one place to another in search of new pasture as the old lands were grazed out, or when it became too dry. And because the change of abode occurred so often, the pioneers never put themselves to the trouble of erecting a decent dwelling. Just like the nomadic Scythians, they lived in simple shepherds' huts, like the huts that the Scythians called *thurgia*.[155]

In this way the half-nomadic, carnivorous hunter-stock farmer came into being—the *migrant farmer*, who right through the course of our pioneer history stood at the outposts of civilization, and during the eighteenth century conquered a large part of South Africa for the white man before the Bantu, who were expanding westward, had advanced very far.

NOTES

1. C 1, Resolutions, 21 February 1657. [Trans. note: The old volumes for Resolutions C 1 through 35 have been withdrawn and are now

available only on microfilm. To lessen confusion I have retained vdM's call numbers for these volumes, although the researcher should be aware that the volumes have been renumbered twice since he did his research.]

2. C 1, Resolutions, 21 February 1657 and Verbatim Copie, (VC) 2, Journal of Cape Governors, p. 304.

3. Leo Fouché, *Die Evolutie van die Trekboer. Lesing gehou voor die Christelike Jongelieden Vereniging* (Pretoria, 1909), p. 3.

4. VC 2, Journal of Cape Governors, 23 December 1658, p. 1433.

5. C 1, Resolutions, 21 February 1657.

6. C 278, Letters Rec., 16 April 1658, p. 1036.

7. VC 2, Journal of Cape Governors, 23 December 1658, p. 1433. [Trans. note: one muid equals approximately three bushels.]

8. C 1319, Letters Disp., J. van Riebeeck to Patria, 5 March 1659, p. 898.

9. Cf. A. J. Du Plessis, "Die Geskiedenis van die Graankultuur in Suid-Afrika: Tydens die Eerste Eeu, 1652–1752" (University of Stellenbosch, M.A. thesis, 1929), p. 111. [Trans. note: The page numbers given by vdM are from the thesis manuscript. This work was later published under the same title, in the series *Annale van die Universiteit van Stellenbosch*, XI (September 1933): 9–127, Cape Town: Nasionale Pers, 1933.]

10. C 2334, Petitions and Instructions, Z. Wagenaer to C. van Quaelbergen, 24 September 1666.

11. C 1319, Letters Disp., J. van Riebeeck to Patria, 5 March 1659, p. 898.

12. C 1, Resolutions, 7 September 1663.

13. C 2334, Petitions and Instructions, Z. Wagenaer to C. van Quaelbergen, 24 September 1666, p. 170.

14. C 1324, Letters Disp., Z. Wagenaer to Patria, 21 November 1663.

15. C 1329, Letters Disp., Z. Wagenaer to J. Maetsuijcker, 24 July 1666.

16. C 1358, Letters Disp., Crudop to Patria, 18 April 1679, p. 111.

17. C 1363, Letters Disp., Simon van der Stel to Patria, 21 March 1681, pp. 719, 775.

18. C 2, Resolutions, 4 October 1664, p. 47.

19. C 1327, Letters Disp., Z. Wagenaer to R. van Goens, 14 January 1665; C 2334, Petitions and Instructions, Z. Wagenaer to C. van Quaelbergen, 24 September 1666.

20. C 2, Resolutions, 4 October 1664, p. 47.

21. Ibid., p. 49.

22. AR, Kol. Arch. 4023, J. Borghorst, Petition, 14 March 1670.

23. C 1352, Letters Disp., J. Bax to Patria, 14 March 1677, p. 430.

24. Cf. Du Plessis, "Geskiedenis van die Graankultuur," p. 111.

25. C 1321, Letters Disp., Z. Wagenaer to Patria, 16 September 1662, p. 148.

26. C 1321, Letters Disp., Z. Wagenaer to Patria, 10 August 1662 and 16 September 1662.

27. C 1324, Letters Disp., Z. Wagenaer to Patria, 21 November 1663, p. 265.

28. C 1321, Letters Disp., Z. Wagenaer to Patria, 30 June 1662, p. 148; C 2334, Petitions and Instructions, 24 September 1666, p. 170; C 1352, Letters Disp., J. Bax to Patria, 14 March 1677, p. 429; C 1360, Letters Disp., Simon van der Stel to Patria, 23 December 1679, p. 389.

29. C 1334, Letters Disp., J. Borghorst to Patria, 19 March 1670, p. 619.

30. C 2270, Original Ordinance Book, 4 March 1678.

31. C 4, Resolutions, 21 April 1679, p. 42.

32. C 1361, Letters Disp., Simon van der Stel to Patria, 31 December 1679, p. 389.

33. C 2, Resolutions, 1670, pp. 339–40.

34. C 1358, Letters Disp., J. Bax to Patria, 18 May 1678, p. 972.

35. C 1358, Letters Disp., Crudop to Patria, 18 April 1679, p. 115.

36. C 2334, Petitions and Instructions, Z. Wagenaer to C. van Quaelbergen, 24 September 1666, p. 167; C 1321, Letters Disp., Z. Wagenaer to Patria, 30 June 1662, p. 84; G. M. Theal, *History and Ethnography of South Africa before 1795* (London, 1909–1910), III(a):66.

37. C 2334, Petitions and Instructions, Z. Wagenaer to Van Quaelbergen, 24 September 1666, p. 176.

38. C 1332, Letters Disp., Borghorst to Patria, 23 March 1669, p. 279.

39. C 1355, Letters Disp., Bax to Patria, 18 May 1678, p. 972.

40. C 2, Resolutions, 1670, p. 338.

41. C 4, Resolutions, 5 August 1679, p. 101, and 23 March 1680, p. 91.

42. H. A. van Rheede, Instructions to Simon van der Stel, 24 March 1686, in G. M. Theal, *Belangrijke Historische Dokumenten over Zuid-Afrika* (Cape Town, 1896, 1911), I:28.

43. C 1381, Letters Disp., Simon van der Stel to Patria, 26 April 1688, p. 44.

44. 1/STB 19/25, Notices, 12 August 1686.

45. C 1378, Letters Disp., Simon van der Stel to Patria, 18 April 1687, p. 652.

46. C 1366, Letters Disp., Simon van der Stel to Patria, 23 April 1682.

47. Ibid.

48. C 1383, Letters Disp., Simon van der Stel to Patria, 26 April 1689, p. 91.

49. Ibid., 15 April 1689, p. 328.

50. C 1366, Letters Disp., Simon van der Stel to Patria, 23 April 1682, p. 45.

51. E. C. Godée Molsbergen, *Reizen in Zuid-Afrika* (The Hague, 1932), IV:296.

52. C 1383, Letters Disp., Simon van der Stel to Patria, 15 April 1689, p. 328.

53. C 1406, Letters Disp., Simon van der Stel to Patria, 30 June 1697, p. 613.

54. C 1406, Letters Disp., Simon van der Stel to Patria, 30 June 1697, p. 614.

55. C 6, Resolutions, 25 April 1686, p. 1.

56. C 2271, Original Ordinance Book, 25 April 1686, new pp. 2–5.

57. Ibid., 22 January 1692, p. 126.

58. C 1, Resolutions, 21 February 1657.

59. C 5, Resolutions, 16 January 1686.

60. C 6, Resolutions, 2 December 1697; 1/STB 1/21 Resolutions, 3 September 1696.

61. C 1, Resolutions, 20 August 1658.

62. C 1320, Letters Disp., J. van Riebeeck to Patria, 9 April 1662, p. 15.

63. C 1343, Letters Disp., Commander to Patria, 10 March 1673, p. 603.

64. AR, Kol. Arch. 3994, Tax rolls for 1682.

65. AR, Kol. Arch. 4004, Tax rolls for 1689.

66. Cape Archives, Tax rolls, J 183, Tax rolls for Stellenbosch [and Drakenstein], 1692.

67. C 2, Resolutions, 5 September 1668, p. 260.

68. Ibid., 5 March 1670, p. 390.

69. C 1340, Letters Disp., Commander to Patria, 19 April 1672, p. 95.

70. C 3, Resolutions, 25 February 1678, p. 440–42.

71. Instructions from Simon van der Stel to Willem Adriaan van der Stel, 30 March 1699, in *Collectanea* (Cape Town: Van Riebeeck Society, 1924), series I, vol. 5, p. 13.

72. Cf. the annual tax rolls in AR, Kol. Arch., nos. 4003–48.

73. Ibid.

74. Cf. Instructions from Simon van der Stel to Willem Adriaan van der Stel, 30 March 1699, in *Collectanea*, pp. 11–24; C 2271, Original

Ordinance Book, 19 October 1691; Tax roll lists in AR, Kol. Arch., nos. 4003–48.

75. C 3, Resolutions, 25 February 1678, p. 440.

76. C 6, Resolutions, 26 February 1688, p. 220.

77. Instructions from Simon van der Stel to Willem Adriaan van der Stel, 30 March 1699, in *Collectanea*, p. 12.

78. Ibid.

79. C 1358, Letters Disp., Simon van der Stel to Patria, 18 April 1679, p. 115.

80. C 2271, Original Ordinance Book, 19 October 1691 [date signed], 22 January 1692 [date published], p. 103; C 6, Resolutions, 19 October 1691, p. 397.

81. C 2271, Original Ordinance Book, 19 October 1691 [date signed], 22 January 1692 [date published], p. 103.

82. C 1319, Letters Disp., Commander to Patria, 9 March 1660, p. 1086.

83. VC 36, Reports of the Commissioners, Frisius, 4 July 1661, p. 141.

84. C 2268, Original Ordinance Book, 2 February 1661, p. 161.

85. C 4, Resolutions, 5 August 1679, pp. 101–2.

86. C 6, Resolutions, 26 February 1688, p. 222; C 2271, Original Ordinance Book, 27 February 1688, p. 54–55.

87. C 6, Resolutions, 19 October 1691; C 2271, Original Ordinance Book, 19 October 1691 [date signed], 22 January 1692 [date published], p. 104.

88. P. E. Roux, *Die Verdedigingstelsel aan die Kaap onder die Hollands-Oosindiese Kompanjie* (Stellenbosch, 1925), pp. 111–24.

89. C 2, Resolutions, 11 July 1673, p. 471; C 2, Resolutions, p. 1744.

90. C 4, Resolutions, 5 August 1679.

91. C 1381, Letters Disp., Simon van der Stel to Patria, 26 April 1688.

92. Theal, *History and Ethnography of South Africa*, II:156.

93. Ibid., II:324.

94. H. A. van Rheede, Instructions to Simon van der Stel, 24 March 1686, in Theal, *Belangrijke Historische Dokumenten*, I:43.

95. C 1381, Letters Disp., Simon van der Stel to Patria, 26 April 1688, p. 39.

96. C 2267, Original Ordinance Book, 31 July 1657, p. 68.

97. C 1, Resolutions, 16 October and 17 October 1657.

98. C 1, Resolutions, 8 November 1657; C 1871, Logbook, 18 November 1657, pp. 304–05.

99. C 1, Resolutions, 30 November 1657.

100. Robert Fruin, *Tien Jaren uit den Tachtigenjarigen Oorlog, 1588–1598* (The Hague: M. Nijhoff, 1906), p. 209. [Trans. note: vdM does not include this volume in any bibliography so I am giving here the edition with the same page number he cites that contains the proverb quoted in the text. The date contained in this footnote in the Afrikaans edition is surely a typographical error, related to the following two footnotes.]

101. C 2267, Original Ordinance Book, 13 September and 24 October 1658; C 1, Resolutions, 24 October 1658.

102. C 2267, Original Ordinance Book, 24 October 1658, p. 110.

103. C 2269, Original Ordinance Book, 25 December 1667, p. 242.

104. C 3, Resolutions, 9 November 1677, p. 356.

105. C 2269, Original Ordinance Book, 25 December 1667, p. 241.

106. C 1325, Letters Disp., Commander to Patria, 15 April 1664, p. 643.

107. C 3, Resolutions, 9 November 1677, p. 356.

108. Ibid., p. 353.

109. C 6, Resolutions, 20 July 1693; Instructions from Simon van der Stel to Willem Adriaan van der Stel, 30 March 1699, in *Collectanea*, p. 15.

110. C 1403 and 1404, Letters Disp., Simon van der Stel to Patria, 1 August 1696, p. 161.

111. C 6, Resolutions, p. 475; C 2272, Original Ordinance Book, 20 July 1693, p. 177.

112. C 2272, Original Ordinance Book, 20 July 1693, p. 175; C 6, Resolutions, p. 451.

113. CJ 3, Criminal and Civil Legal Rolls, p. 4.

114. Ibid., p. 52.

115. Ibid., p. 37.

116. C 21, Resolutions, 13 January 1693, p. 470.

117. Instructions from Simon van der Stel to Willem Adriaan van der Stel, 30 March 1699, in *Collectanea*, p. 13.

118. Ibid.

119. Ibid.

120. See the annual tax rolls in AR, Kol. Arch., nos. 4003–48.

121. C 6, Resolutions, 1 February 1691; C 681, Original Ordinance Book, 13 February 1696.

122. Cf. C 2271, Original Ordinance Book, 19 October 1691, p. 102.

123. C 6, Resolutions, 18 April 1687, p. 649. [Trans. note: For a description of this target practice at a wooden parrot, see G. M. Theal, ed., *Chronicles of the Cape Commanders* (Cape Town, 1882), p. 266.]

124. C 1341, Letters Disp., Commander to Patria, 19 April 1672, p. 118; C 2, Resolutions, 11 July 1673, p. 471.

125. C 3, Resolutions, 26 July 1677, pp. 319–20.

126. C 6, Resolutions, 18 April 1687, p. 649.

127. C 2270, Original Ordinance Book, 8 April 1680, p. 409.

128. C 2272, Original Ordinance Book, 21 February 1692, p. 118.

129. C 1376, Letters Disp., Simon van der Stel to Patria, 12 April 1686, p. 285.

130. C 2270, Original Ordinance Book, 8 April 1680, p. 409.

131. Cf. RLR 1, Old Gamehunters Books, I, October 1687 to October 1712. [Trans. note: According to the inventory for the Receiver of Land Revenue (RLR) in the Cape Archives Depot, the "Oude Wildskutte Boeke [Old Gamehunters Books], 1680–1686, have not been preserved." Old S.G. 1 is now RLR 1, and covers the dates October 1687 to October 1712.]

132. Instructions from Simon van der Stel to Willem Adriaan van der Stel, 30 March 1699, in *Collectanea*, p. 13.

133. C 1381, Letters Disp., Simon van der Stel to Patria, 26 April 1688, p. 44.

134. Instructions from Simon van der Stel to Willem Adriaan van der Stel, 30 March 1699, in *Collectanea*, p. 13.

135. C 6, Resolutions, 19 October 1691, p. 400.

136. C 2271, Original Ordinance Book, 19 October 1691 [date signed], 22 January 1692 [date published], p. 104.

137. VC 15, Journal of Cape Governors, 31 July 1700, p. 763.

138. C 1418, Letters Disp., 1 March 1700, p. 690.

139. VC 15, Journal of Cape Governors, 31 July 1700, p. 763.

140. C 1418, Letters Disp., 1 March 1700, pp. 689–90.

141. VC 15, Journal of Cape Governors, 31 July 1700, p. 763.

142. P. Kolbe, *Nauwkeurige en Uitvoeringe Beschrijving van de Kaap de Goede Hoop* (Amsterdam, 1726), I:127, 130; F. Valentyn, *Oud en Nieuw Oost Indiën* (Amsterdam, 1727), V(II):38.

143. C 2274, Original Ordinance Book, January 1705, p. 336.

144. C 1435, Letters Disp., 28 March 1705, p. 509.

145. Kolbe, *Beschrijving*, I:77.

146. Valentyn, *Oud en Nieuw Oost Indiën*, V(II):47.

147. C 1435, Letters Disp., 28 March 1705, p. 509.

148. A. Bogaerts, *Historische Reizen door d' Oostersche Deelen van Azie* (Amsterdam, 1711), p. 509.

149. C 1435, Letters Disp., 28 March 1705, p. 509.

150. Kolbe, *Beschrijving*, I:124–31.

151. Valentyn, *Oud en Nieuw Oost Indiën*, V(II):38.
152. Ibid., V(II):37.
153. Kolbe, *Beschrijving*, I:126.
154. Ibid., I:125.
155. Ibid., I:124.

II

The Quitrent Farm System

THE ORIGIN OF PRIVATE
GRAZING RIGHTS

The System of Communal Grazing

The first free burghers at the Cape received in full ownership as much land as they could bring under cultivation within three years. The Company fairly faithfully pursued this principle of allocating private land intended for agriculture until the end of the seventeenth century. Such land was later surveyed and mapped, and the owners could lease it, sell it or bequeath it. The free burghers, however, kept livestock as well as practiced agriculture. Naturally they could not maintain their livestock on their small individual farms, yet the government had made no provision for private grazing. In imitation of the common pasturage system, so familiar in Europe and also in vogue in certain parts of Holland, it was declared that the colonists could use the "entire country" as grazing land.[1]

In February 1678 the government granted freedom to Henning Huising and Claas Gerrits, two of the Company's cattleherds who possessed no freehold property, and allowed them to begin stock farming on their own account in the vicinity of the Steenberg.[2] In August of that same year Cornelis Stevensz appeared with a petition to allow his livestock to graze next to the Wynkelder. This petition was also granted, because the Council of Policy determined that such permission would not be detrimental to the Company, which itself continued to raise livestock. Stevensz was warned, however, that he would have to find another place to graze his stock if the Company later needed the fields for its own herds.[3] It seems clear that the Company still considered itself master of the grazing lands. The intention was not to give away permanent grazing rights, but merely to make a temporary concession that could be arbitrarily revoked at any time.

It is doubtful whether these farmers intended to try and acquire private grazing rights on particular pieces of land with their petitions. Rather, they evidently asked for special permission to graze their livestock in the vicinity of the Steenberg only because this region was somewhat remote and lay outside the recognized grazing area. It is also doubtful whether the fact that grazing rights in these cases were specially granted, while the right to let livestock graze on undistributed crown land was regarded as something of a matter of course, can be considered as an important deviation from the original system of common grazing. Henning Huising and his partner received the right to graze their livestock "in the furthermost corner of the Steenberg, stretching seaward and near the Caap Fals [Cape Hangklip]."[4] Cornelis Stevensz was allowed to settle "behind the Steenberg, at or near the place called the Wijnkelder."[5] The farmers were thus given grazing rights in the same region, but in both cases only a vague indication of the locality was given. There is no mention of two clearly defined grazing areas that were definitely separated one from the other. It is therefore questionable to assume that we are already dealing here with the beginnings of a system of private grazing rights.

The development of private stock farming as an independent occupation, existing separately from agriculture, made a system of private grazing rights very desirable during the last years of the seventeenth century. If the Cape government had continued to grant grazing licenses, which had begun under Johan Bax toward the end of the 1670s, undoubtedly such a system would already have developed in a spontaneous manner before the end of the seventeenth century. Bax was followed, however, by a commander who had little respect for a farmer who did not know how to handle a plow, and who was resolved to suppress independent stock farming.

According to Simon van der Stel's conceptions of colonization, expansion had to be based on densely populated agricultural settlements. Where the ground was suitable for cultivation, the colonists had to settle near to one another on small freehold farms. They could practice stock farming as a secondary occupation, but primarily they had to cultivate the land. The well-cultivated freehold farm therefore had to remain the absolute starting point for their stock farming. The undistributed crown land in the immediate vicinity of the settlement could be used communally by the colonists. The right to use the Company's grazing land followed automatically from the possession of freehold property, and the right remained linked to this possession. Simon van der Stel would not recognize individual grazing rights that were not based on the proprietary right. This is clear from the ordinance specifying that no one would be allowed to keep livestock unless he was "housed and propertied," and that farmers must return home every evening with their livestock after they had spent the day grazing them in the fields.[6]

Simon van der Stel thus still planned to maintain the system of common grazing. In 1687 he designated a piece of land as a communal pasture at the request of Stellenbosch farmers,[7] and this ground could only be used by them. The grazing areas of the Cape and Stellenbosch would be kept strictly separate and livestock from one district was not allowed into the other. Stellenbosch farmers were forbidden from even keeping the Cape farmers' livestock among their own herds for payment.[8]

A few years later the grazing land in both districts became too limited. To make further emigration toward Stellenbosch and Drakenstein possible, the governor expanded the district's borders in 1691, so that the farthest end of the Paarl Mountain and the Babylon Tower were included.[9] In order to provide the Cape colonists with more pasture the following regulation was enacted:

> [It is] hereby thought proper to permit and allow, and with this we do permit and allow, them to graze their livestock outside the jurisdiction, and borders of Stellenbosch and Drakenstein, to wit, from Paarl in a slanting straight line to the Cape of Good Hope, and once again from the old ford at Stellenbosch measured in a slanting straight line to Table Mountain, provided they shall have received the prior written consent of the Honorable Lord Governor and that it shall be properly registered with the colonial secretary.[10]

The purpose of this regulation, which allowed grazing licenses to be issued to individual farmers, does not appear to have been a deviation from the principle of common grazing and an attempt to substitute a system of private grazing rights in its stead. Apparently the governor's intention was only to extend the communal grazing area of the Cape district. The reason he demanded that the farmers obtain his written consent before they sent their livestock to the newly added pasture is obvious. The area lay so far from the Cape that the farmers would find it impossible to use it for grazing if they had to return home every evening with their livestock as the law required. In addition, such licenses were necessary in order to exercise control over the migrant farmers who traveled about with livestock in the interior without possessing and cultivating land.

Grazing Licenses and Livestock Posts

It is not known if during the last decade of the seventeenth century permission had already been granted to migrate inland with live-stock and to establish posts there. Naturally, it could have happened but there is no mention of it in the contemporary sources. Only from 1703 are grazing licenses also registered here and there among the hunting licenses in the Old Gamehunters Books.

It is evident when one compares these first grazing licenses with the yearly returns that they were chiefly used by persons already settled on freehold farms. Furthermore, it is noteworthy that the period of validity often varied on the earliest grazing licenses. Up to 1705 a total of fifteen licenses were issued. Of those, seven were valid for three months, one for four months, two for six months, one for nine months, and two for a year. In two cases the duration of the license is not stated.

Hence it appears that during the period when stock farming was still essentially an adjunct to agriculture, livestock posts were merely supplements to the freehold farm. Freehold land always remained the starting point for individual farming. When the situation demanded it, the government simply granted permission to established farmers to use the open country in the interior for as long as it was necessary. The first grazing licenses were consequently sporadic concessions to provide for temporary and passing needs. They were not particularly intended for extensive stock farmers who possessed no freehold land and therefore had a permanent need for grazing land. By the beginning of the eighteenth century such farmers already existed but they still wandered about the interior on their own. It was not till a few years later that they began to take out grazing licenses. Admittedly, with the increase in extensive stock farming there gradually developed out of the system of grazing licenses a fixed form of land tenure for farmers without freehold land, but this occurred in a spontaneous way. The system of grazing licenses was not originally drawn up with this purpose in mind.

Furthermore, it is important to point out that in the earliest grazing licenses no precise locations are indicated where the farmers "had to settle and graze" their livestock. They all contain only a vague indication of a locality. In 1704, for example, all the licenses granted permission to graze in the dunes behind the "Graauwe heuvel (burial mound)," but none of them named exact sites. In 1705 grazing rights were allotted on the same spot to five out of the eight farmers who took out licenses. The following year thirty-two out of the forty-four farmers to whom grazing licenses were issued received the right to let their livestock graze "in Groen-kloof."[11] Consequently, since grazing rights in the same vicinity were granted year after year to different farmers, without each one being allocated a separate grazing area, it was obviously not the governor's intention to grant individual license holders private grazing rights on separate pieces of land. Clearly the license holders, following time-honored tradition, had to use this pasture in common. Originally therefore, the system of grazing licenses only supplemented the system of communal pasture.

The Regular Occupation of Fixed Sites

As a general rule, grazing licenses apparently did not promote the settled occupation of particular areas during the first years of the eighteenth century. Indeed, for a farmer who owned a freehold farm and merely drove his livestock temporarily into the open country, there was little inducement to limit the livestock to a fixed site. His grazing license did not bind him to a specific spot and he therefore apparently had the right to migrate with his livestock from one place to another as pasture or water became scarce or if too many other farmers had entered his neighborhood. There was undoubtedly much use made of this privilege. Valentyn says that many farmers at the Cape had four, eight, sixteen, and more livestock posts deep in the interior,[12] although in his time, according to the Old Gamehunters Books, no farmers had more than one grazing license registered in their name. Consequently, if we accept Valentyn's word, we must assume that the farmers simply allowed their livestock to wander from one temporary livestock post to another, and that the number of temporary livestock posts a license holder used depended on his own personal discretion.

This conjecture is strengthened by the description of a livestock post that Willem Adriaan van der Stel, the governor who issued the first grazing licenses, gave in his "Short Testimony":

> And so to consider the fifteen present livestock posts, mentioned in the third article above. It must be recognized that prior to this time, livestock posts at the Cape were called boundary fences, or dividers, and were constructed in the open field with all sorts of branches from bushes and shrubs, and scrap lumber. These were placed then around an unoccupied site, or were placed or stuck in the ground. And then here or there men grazed their livestock in these places in order to have the best grazing, or out of need for water. In order also to drive their livestock at night to the same open temporary places for security against various wild animals. A hut of straw or branches was erected nearby as well, of such a circumference that two to three cattle herders could sleep together at night. So it is easy to understand that the creation of a few or many such places, or livestock posts, ultimately depended on the needs of the livestock attendants.[13]

The tendentious character of this source does not mean we have to doubt the reliability of the description. In their "Counter Testimony," Adam Tas and van der Heiden—the governor's opponents—refer to this description of a stock post and cite it as authoritative.[14]

Nevertheless, with the passage of time livestock posts were set up that were more or less habitually occupied. And, closely linked to this development, stock farmers possessing no freehold land gradually began to make use of grazing licenses. Already by 1706 there were among the license holders for that year a number of farmers who owned no permanent places of residence and farmed on the basis of temporary grazing licenses. Every year this class of license holder grew and by the end of the eighteenth century the vast majority of loan farm tenants were people who possessed no freehold land. Accordingly, as grazing licenses became in time the sole basis of extensive stock farming, the use of livestock posts became more permanent. Now the livestock post was no longer merely a temporary extension of the freehold farm, but it had to provide for a permanent need for pasture, since the stock farmer had to keep his livestock continually in the interior. Consequently, the practice of renewing grazing licenses developed very shortly thereafter. In 1706, when grazing licenses were still mostly issued for six months only, we find for example that sixteen of the forty-four grazing licenses were renewed again in the same year.[15]

Even farmers who farmed throughout the year on regularly renewed, but temporary, grazing licenses, would of course still migrate frequently with their livestock. And gradually there would develop among them a desire to have fixed family abodes to which they could always return. Already the pliable system of grazing licenses had resulted in specified sites being more or less regularly occupied, and in this manner the solitary livestock post became in the course of time the permanent home of many farm families.

Another factor facilitating the permanent occupation of livestock posts was the fact that as time went on farmers had begun to build on them. Even before the farmers began making a practice of taking out grazing licenses, they constructed "houses, cages, and pens" on the livestock posts they set up on their own in the interior.[16] The migrant farmer, who possessed no freehold land, naturally had a greater need for this than the settled agricultural farmer, who had a permanent residence in the inhabited areas. As soon as a farmer made improvements on a livestock post, there was more incentive to return there regularly, even though he might temporarily abandon it now and then.

Migrant farmers—and evidently especially those who possessed no freehold land—soon displayed a tendency as well to till the land around their livestock posts and at least to harvest enough wheat for their own bread. The right to sow grain on a livestock post did not result naturally from the grazing license. Initially, if a farmer wanted to cultivate the land, he had to request special permission beforehand. Now and then it was expressly granted in the grazing license and sometimes it even stated how many sacks of grain might be sown. [17] But as more and more farmers who owned no freehold land began to farm on livestock posts, the right to cultivate on these posts became a matter of course, and it was no longer referred to specifically in the grazing licenses. Farmers wanting to sow and plant could no longer wander about as much, however, and thus the cultivation of livestock posts made their occupation more permanent. It is also noteworthy that while only a vague indication of the locality was given in the common grazing licenses, a specific spot was generally designated on the licenses in which permission was also granted to sow grain. Sometimes the name of the place was specified; sometimes the location was fairly precisely indicated. [18]

Also working against the mobility of the license holders was the geographical fact that permanent open bodies of water could not be found everywhere in the interior. When a farmer discovered good water, that gave him sufficient reason to set up a permanent post at that spot to which he could regularly return. This factor became of greater importance as the stock farmers migrated deeper and deeper inland.

As farmers under the grazing license system gradually occupied fixed livestock posts on a permanent basis, the practice of granting grazing rights to different farmers in the same vaguely designated vicinity slowly disappeared. Little by little it became the established custom to allocate to every individual license holder a definite spot to which he could go with his livestock to graze. The location of such a spot was described relatively inaccurately right up to the end of the eighteenth century, but very early on it became a tradition to record on the license a name for the specific place by which it would be known in the future to the government and the public. In this way it was possible later to locate where a specific license holder had settled with his livestock, and also to prevent the government from unlawfully granting a farmer rights to a livestock post that was still occupied by someone else.

A farmer desiring grazing rights for his livestock normally picked out the spot himself where he wanted to construct his homestead. It later became the practice that as soon as someone found an acceptable spot for a homestead he erected a marker there to make known his intention to other farmers. After that he requested permission from the governor to settle there with his livestock. The place where the marker stood—or where the first homestead was constructed—was then regarded as the point of departure for his grazing rights. For that reason it later become an unwritten law that a license holder had to construct his homestead at his marker. And if he later moved his homestead, he had to make that fact known to the magistrate. That change was then recorded, so that it was possible to find the spot where the original marker had stood, in case some uncertainty later developed in connection with the grazing rights of the license holder concerned.[19]

Private Grazing Rights

It is understandable that as soon as a farmer began to till and build on a livestock post, he would become eager to obtain private grazing rights around that site as well. If he did not have the right to keep other farmers out of the vicinity of his homestead, intruders could, by allowing their livestock to consume the pasture, oblige him to migrate to another area, even perhaps while his crop was still standing in his fields. There he then would have to build a new house or live in his wagon tent. Moreover, private right of use to a specific piece of land was very desirable for the stock farmer. Through it he could acquire control over the water on the property and the use of a large piece of grazing land frequently depended on this. Consequently, as the livestock post became the sole basis for stock farming, it became necessary to replace the obsolete system of common grazing with one of private grazing rights.

This development was inevitable as soon as the government allowed license holders to settle permanently on specific sites, to build houses and cattle pens there, and to till the soil. Indeed, the right to occupy a livestock post and to cultivate, would be worthless to a farmer if the government did not grant him private grazing rights all around his place. The government soon made it clear as well that it would not permit other farmers to bother an established license holder on the grazing land he required. As early as 1708 the government began implementing this principle. On 24 October of that year a notation was entered on the grazing license of B. P. Blom that whereas the widow Elbertzn had complained that he was "lying too close" to her, the license would not be renewed.[20] With that the principle of private grazing rights was formally adopted.

Certainly not all license holders had permanently occupied fixed livestock posts by this time, and perhaps not everyone had yet manifested the desire to obtain private grazing rights. Many license holders still owned freehold farms in agricultural districts and they therefore had less interest in permanent livestock posts than did stock farmers without freehold land. In addition, as long as there was still abundant, unoccupied land, it was a great temptation for any license holder to travel about for game and green pastures. But as the population in the border districts grew and conditions stabilized, it became all the more necessary to acquire permanent living places and to obtain private grazing rights to the surrounding pasture.

The development of private grazing rights at the Cape was closely linked with the manifestation of extensive stock farming that had gradually replaced agriculture as the basis of economic life in the colony by the beginning of the eighteenth century. The system of common grazing perhaps could have continued to exist if the climate, soil conditions, transportation facilities, and marketing opportunities for agricultural products had made land colonization on the basis of group settlement possible. Such an environment would have promoted agriculture first, with stock farming practiced only as a secondary occupation. In reality, however, it was not the little group of agriculturists who together had formed a small village, but the isolated, patriarchal stock-farming family, sometimes strengthened by a share-cropper or two, that was the colonizing unit leading the expansion into the interior from the beginning of the eighteenth century. And when stock farming freed the stock farmer-hunter from agriculture and he migrated alone into the interior, the common grazing system reached the end of its usefulness and another form of land ownership, which more effectively suited the needs of the extensive stock farmer, became desirable.

When private stock farming finally began on a large scale, it was only possible under South African conditions to use the pasture communally if the leaseholders would also farm communally and travel about in groups. The individualistic South African colonists were not inclined, however, to farm like the nomadic Khoikhoi. The communal idea, as found among nomadic peoples, was foreign to our farmers, with their strongly developed ideas about private property. They wanted to keep their farms separate, and for that reason they had to live dispersed from one another and use the land on the basis of individual grazing rights—regardless of whether their grazing areas were allocated through the government or separated by mutual consent.

In the entire organic growth of the quitrent farm system the individualism of the Cape colonists shines clearly through. Already by Simon van der Stel's time, Jan Vosloo, whose livestock herds were situated across the Berg River below the Lemietberg, informed Henning Huising, "Either you or I have to leave from here, because we are bad for one another, and already our sheepherders have had each other by the ears."[21] In the eighteenth century the farmers viewed their affairs in precisely the same light and the only solution was to divide up the pasture as Lot and Abraham did. Undoubtedly the development of private grazing rights under the grazing license system is attributable in the first place to the conscious endeavors of the colonists—and not necessarily to the government's influence.

With the formation of the winter farm system in the Karoo, which developed as a supplement to the quitrent farm system, we notice also the colonists' eagerness to farm their own ground and be the master thereof. Every winter the Karoo was visited for a few months by farmers from the more highly elevated border areas, and initially they used the land in common. With the passage of time, however, the farmers picked out individual farms for themselves and began to return to those places regularly. They mutually recognized one another's claims to these "winter farms" and eventually the government had to recognize them as well, although it had repeatedly insisted in the past that the Karoo could not be divided but had to be used as communal winter pasture.

Only in the arid "Agterveld [back country]," where the farmers of the northwest had to migrate with their livestock during the summer as a result of drought, were they content to let their livestock graze communally on the land. Every year the rains fell there in a different place, and as a consequence private grazing rights to specific pieces of ground were nearly worthless. A farmer had to have the right to migrate where the rain fell, otherwise he could not keep his livestock alive in the "Agterveld." This climatological factor explains why the farmers in the northwest wished to graze the open country on the basis of the "grass license," while they purposefully sought to acquire private grazing rights on individual farms elsewhere in the colony where soil conditions permitted it.[22]

THE SHAPE AND SIZE
OF THE LOAN FARM

Unlimited Grazing Rights

There was still no notion of a specified amount of pasture in terms of morgen when grazing licenses were issued at the beginning of the eighteenth century. From Willem Adriaan van der Stel's description of a livestock post—which we have already cited—it is clear that a license holder could freely use as much land as he needed. We must draw the same conclusion on the basis of the contents of the grazing license. This document granted the holder the right "to settle and graze" his livestock in a certain vicinity or at a specified place with no further description of the grazing area. The only restriction on these vaguely written grazing rights was the stipulation "that they should take care not to disturb anyone already settled there."*

This liberal position of the government with regard to the temporary use of undistributed crown land ought to come as no surprise. In the beginning the government could not limit the individual license holder's grazing area to a certain number of morgen, because there was not a separate grazing area allotted to each one; the first license holders apparently had to use the pasture in the old way: communally. When private grazing rights developed in the course of time, it was not necessary for the government to concern itself about the amount of grazing land that a license holder

*[Trans. note: No citation is given for these quotations in the Afrikaans edition.]

used. Pasture was so plentiful that there was more than enough for every farmer. In addition, the grazing licenses were revokable permits, from which the Company received no income. The temporary use of grazing land, of which the Company remained the owner, could therefore not cause it any harm.

But although the government's point of departure was that each license holder must have *enough* grazing land for his livestock, the principle was also maintained that *every* farmer had the right to make a claim on it. For that reason licenses were issued to anyone who applied for them. And later, when the government received an income from them, there was yet further incentive to approve requests for grazing licenses. This circumstance now created the necessity of placing certain restrictions on the unbounded grazing rights of individual license holders, and the need became urgent when the farmers gradually began to develop exclusive grazing rights around specific points. Under a system of private grazing rights, there would inevitably be conflicts over pasture unless the different license holders' grazing areas were precisely described.

Nevertheless, for years the government did nothing to avert such conflicts over grazing land. It is sometimes maintained that a preliminary investigation was set up before new grazing licenses (later called loan farm leases) were granted. The government supposedly sent out commissioners to determine if the distribution of a new farm encroached on the rights holders of established loan farm leases.[23] However, there is no indication in the reports of the deputy heemraden, nor in the minutes of the landdrost and heemraden, that anything like this occurred during the first eighty years of the eighteenth century, when the loan farm system gradually took fixed shape. In reality, preliminary investigations for the granting of new grazing licenses would have been too cumbersome and too costly at a time when landdrosts were still scarce. Furthermore, during the period of free expansion the government could depend on the majority of farmers not placing their landmarks too near the property of the established loan farm leaseholders. The individualistic migrant farmer found it in his own interest to live as far as possible from his nearest neighbor. What is more, the distribution of permanent pools of water inevitably meant that in most cases homesteads could not be established too near to one another. Considering therefore that conflicts over pasture would only occur now and then, it was not necessary in every case to do a preliminary investigation.

During the last quarter of the eighteenth century colonial expansion was temporarily arrested by encounters with the Bantu on the eastern border and with the Bushmen on the northern border. It thus became more necessary than ever to distribute land by leasing it in between existing farms. Naturally this rapidly increased the possibility that permanent license holders would be disadvantaged by new intruders. Probably for this reason the government, at the time of the formation of Graaff-Reinet, stipulated in the instructions for the landdrost and heemraden of the new district that when a new farm was petitioned for on lease, an investigation must take place "to determine whether such place can be distributed without really harming anyone." A written statement to that effect then had to be submitted to the governor before he would allocate the loan farm in question.[24]

Later a preliminary investigation prior to the distribution of new loan farms was required in *all* districts.[25] And apparently such investigations were a regular occurrence after that. At times the landdrost and heemraden instituted the necessary investigation; sometimes it was conducted by the fieldcornet of the area concerned, assisted by two impartial witnesses.[26] By this time, however, the loan farm system had already taken a definite form. It is important, therefore, to understand that during the first years of the eighteenth century, when this form of land ownership had still not taken a definitive shape, the governor did not require an investigation before he issued new grazing licenses.

Instead, he permitted the license holders to settle where they chose with no further ado. In many cases the governor did not even have an approximate idea of where the requested spot lay. The farmer perhaps described the farm as "Leliefontein across the Gamtoos River" or "Driefontein in the Roggeveld." The governor had no idea how far this place lay from other license holders across the Gamtoos River or in the Roggeveld. He did, however, consider the possibility that the allocation of the new loan farm could harm established license holders and for that reason he granted the farm conditionally. The leaseholder could use it "provided he would not be a bother to someone already herding there." This stipulation appeared in all loan farm leases. Apart from that the Company remained owner of the ground and it had, therefore, the right to withdraw new loan farms if it appeared that their allocation infringed on the rights of old loan farm leaseholders.

In actual practice this occurred frequently. If the occupant of an old loan farm complained that the issuance of a new farm harmed him, the deputy heemraden were sent to investigate the complaint and to deliver a written report about it. They then recommended, according to circumstances, that the new farm be withdrawn, that it might remain occupied in loan, or that the grazing land must be divided in a specified manner between the parties concerned.[27] The government generally executed such recommendations without further investigation.

The Minimum Distance between Loan Farms

The question now arises as to what criterion the government used when judging whether or not the distribution of a new loan farm harmed an established loan farm holder. We find the answer in the reports of the deputy heemraden, who were sent out from time to time to investigate objections to the granting of new leases.

From such reports it is clear that the distance between the homesteads concerned was used from the beginning as an important criterion for judging what amount of land was sufficient for successful farms. In addition, it appears that in most cases where the homesteads concerned lay at a distance of one hour's walk or more from each other, the deputies regarded the intervening space as sufficient. If the farms were situated nearer than an hour's walk from each other, however, then the intervening space was usually considered to be insufficient. It is apparent therefore that from the beginning of the eighteenth century an hour's walk was regarded as the natural minimum space between loan farms. The distance between homesteads, however, was not the only, and also not always the most important, criterion when a decision was taken as to whether a new loan farm had to be withdrawn. The following examples will illustrate this.

In 1779 Anthony Rink complained that Pieter Rossouw, by leasing a farm nearby his, harmed him "as much for pasture as for water" for his livestock. The landdrost investigated the matter and, having closely inspected the location of the pasture and water, found that the place of the above-named Rossouw was plus or minus one hour's walk from the residence of the plaintiff, and consequently that as far as this was concerned the distance was quite far enough. However, the poor condition of the pasture on this side, consisting only of brambles and bushes, and without any suitable capacity for grazing nearby the farm in question, must be considered. Furthermore the plaintiff Rink's farm is enclosed on the other sides by mountains as well as two other adjacent farms. The farm has no water, but merely a small stream, which usually dries up in the summer, so that when Rink requires it, he is not only forced to water his livestock from the river where the farm of the defendant is situated, but even has to transport his own drinking water, and he would find it impossible to exist without such water.[28]

For these reasons it was recommended that Rossouw's farm be withdrawn.

A similar case occurred a few years before. A certain Landman raised an objection to the granting of a fixed farm by lease to one van der Bank because, Landman complained, he was harmed by it. The farm in question lay in a plain behind a hill to the northeast of Landman. It was a half hour from Landman's farm up to the hill "where there was nothing other than rocks and shrub, and no grass or pasture for grazing was growing." From the hill, "where the first pasture for Landman's livestock began," it was also a half hour in the same direction up to the farm that was granted to van der Bank. The two farms therefore lay an hour from one another, but in spite of this fact the deputies arrived at the conclusion that if van der Bank continued to occupy the newly granted farm, it would deprive Landman "not merely of the principal grazing land but also of all further expansion," considering that to the north of his farm were cliffs without pasture and to the south there were rocky ridges.[29]

In a case in 1770, where Pieter du Plessis raised an objection to the issuing of a lease to Jacobus Pinard, it was found that the loan farms lay a distance of more than an hour and a half from one another: "So that if the above-mentioned Pinard could make use of the pasture alongside the ravines and hills lying to the north, as he has testified he could do, then du Plessis would retain sufficient space and pasture for his livestock southward up to his farm."[30] In this case, therefore, it was not taken for granted that an intervening space of an hour and a half was large enough.

The deputies further agreed to fix a boundary between the two farmers. The older license holder retained more than three quarters of the land and the new license holder had to compensate for the loss of his rightful half by grazing his livestock in an opposite direction.

Even by the end of the eighteenth century an intervening space of an hour between farms was still not considered as sufficient in all cases. In 1799 a field-cornet and some witnesses, who had to investigate whether Pretoriuskraal (evidently an old livestock post) could be granted to A. C. Greyling, declared that the farm was "absolutely insufficient," although it was situated an hour and two minutes from the nearest neighbor.[31]

On the other hand it sometimes happened that an intervening space of *less* than an hour was regarded as adequate by the deputies.

In 1756 Fredrick Zele, the occupant of the loan farm "Die Uijtvlugt" in Outeniqualand, complained that the farms Diepte Kloof and Klynefontein, requested by ordinance by Hendrik Plooy and Hendrik van der Wath, respectively, "were lying too near to him." Landdrost Horak investigated the matter and found "that the Diepte Kloof lay a full hour's walk from the farm Uijtvlugt in the north, leaving the extension to the west and south uncertain." Under these circumstances the landdrost concluded "that all three farms could remain."[32]

Another interesting case arose in 1776. Johannes van Aarden plowed a piece of land nearby the homestead of his loan farm, "de Knollefontein," and cleared a piece of land that bordered it. Lucas Visagie raised an objection to this by virtue of the fact that it "would seriously disadvantage him in the free movement and grazing of livestock on his farm." Consequently, he demanded that the ground must lie fallow. Van Aarden, however, argued that the ground was part of his loan farm. He declared, moreover, that Visagie had enough pasture for his livestock. The deputies who investigated the matter found "that the piece of plowed land in question, the outermost end of which was found to be upwards of a little more than a quarter hour from the aforementioned Knollefontein and marker, properly ought to belong to the loan farm, but that the remaining parts between the aforementioned land and that of Visagie ought to be left alone and never cultivated, so that they are available for the freedom and movement of the livestock on Visagie's farm, as well as for common grazing."[33] Therefore, van Aarden received exclusive right to use merely a quarter of an hour of the ground in the direction in which Visagie lived.

This was not an extreme example. In 1796 the Bokkeveld field-cornet mentioned two loan farms lying 1,100 yards, roughly ten minutes' walk, from each other.[34] If the intervening space was divided equally between the occupants of the two farms, each one had five minutes of field at his disposal in that direction from his homestead.

While investigating complaints against the distribution of new loan farms, the deputies had to consider not only the distance between the complainant's farm and the new loan farm. The condition of the water and the nature of the pasture also had to be considered, as well as whether there were other neighbors in the vicinity. The cardinal principle guiding the deputies appears to have been that a leaseholder had the right to the use of as much grazing land as he needed. Consequently, what decided the matter was not distance but the sufficiency of the complainant's farm. If a newer license holder depended on the use of land without which an established license holder could not do without, the issuance of the new license was recommended without exception, no matter what the distance was between the concerned farms.

Evidently it was generally accepted that a half hour's walk under normal circumstances was a reasonable intervening space between loan farms but the judging of the sufficiency of farms was dealt with on its own merits in every special case. What was permissible as the minimum intervening space in a specific case depended on the local circumstances. The principle of a minimum intervening space of an hour was thus not maintained as an unbreakable rule. On the other hand it was also not departed from too often, evidently because it was a more or less natural distance and not an arbitrary one (fig. 1).

The principle that every loan farm occupant must have enough pasture for his livestock even gave rise in some instances to a situation whereby the system of private grazing rights that the loan

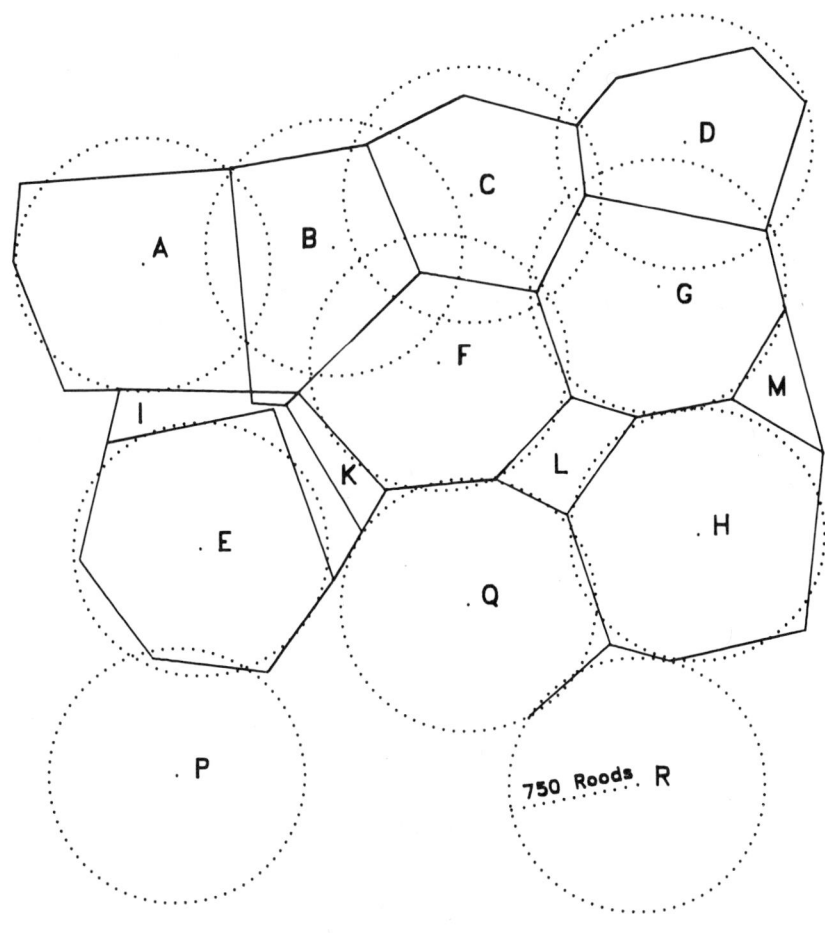

Fig. 1. This map, which was made in 1826, gives the location of eleven loan farms in the field-cornetcy Vetterivier, district of Swellendam: Jonkersfontein (A), Spiegelrivier (B), Hollebak (C), Vergenoegd (D), Karnemelksrivier (E), Kraggakraal (F), Vetterivier (G), Kweekkraal (H), Koeyga (P), Melkhoutekraal (Q), Jan Pienaars Rivier (R). Perpetual quitrent was requested for all the farms, but the last three had not yet been surveyed. I, K, L, and M represent pieces of crown land that were being sought by various farmers.

farm lease introduced was maintained hand in hand with a system of common grazing. This principle is beautifully illustrated by the following examples.

In 1770 J. B. Hoffman owned the loan farm Donkerhoek, that happened to lie just a half hour from Brakkefontein, the loan farm of a certain de Vos. Donkerhoek could not be used for livestock during the winter because of the cold. Accordingly, Hoffman, in imitation of the practice of the previous owners of the farm, drove his livestock across the river during this time of the year and allowed them to graze between the loan farms of de Vos and a certain Morkel. Because of the distance, this ground could not be considered as part of Hoffman's loan farm but for the same reason de Vos and Morkel could also not lay claim to it. Not one of the three parties, therefore, possessed exclusive grazing rights to this piece of land on the strength of their loan farm leases. Nevertheless, Hoffman tried to acquire sole grazing rights by leasing the piece of land in question. The homestead of the new loan farm lay an hour's distance from the farms of Morkel and de Vos, but in spite of this they raised an objection to the issuance of the new farm.

The deputies then went to inspect "the location and distance of the aforesaid farms one from the other, together with the paths out for the livestock in addition to the field there in the neighborhood," and came to the conclusion that these three farms had need of the pasture on the one side of the river in the summer and on the other side in the winter. Consequently, if Hoffman kept the new farm in loan and thereby possess sole grazing rights to it, it would impede Morkel as well as de Vos "in moving their livestock freely across the land and across the river." Accordingly, it was recommended that the farm be withdrawn. During the winter, however, when he had need of the pasture, Hoffman's livestock had to be allowed to graze there. For this purpose he could use the cattle pens he had erected there. For the rest, the farm had to serve as communal grazing land for all three parties. And since de Vos and Morkel had need of the pasture on the other side of the river again in the summer, specific grazing areas were assigned to them there.[35]

The case of Willem Landman and Gideon Joubert Janse vs. Dirk van der Bank illustrates this point as well. Formerly, the three parties had communally used a piece of land that not one of them could consider as part of his own loan farm. Van der Bank then acquired exclusive grazing rights to the pasture by lease; but the deputies found that this would not be desirable, since van der Bank then "had sole use of one large open field that had formerly been used by all three together to graze their livestock."[36]

Although under the loan farm system there developed comparatively early a recognized principle with respect to the minimum-space between homesteads, in practice there was never a stipulated rule followed that prescribed *how far* apart from one another the homesteads of loan farms had to be built. For that matter, it was also impossible to claim that farms had to be laid out at a fixed distance from one another, because the first pioneers depended on the use of permanent natural waters that were not distributed across the country according to some or other geometrical principle. Consequently, land under the loan farm system was also not divided systematically, but the farmers themselves selected their farms and set up their homesteads on sites that appeared environmentally the most promising.

There was nowhere even a rule set down or a principle adopted that defined the maximum space that should be allowed between loan farms. Anyone could select a farm for himself wherever he wanted, no matter how far from his nearest neighbor. Considering, however, that homesteads were permitted up to an hour's distance from each other, a third would be erected between two homesteads if they lay two hours or more from one another. Theoretically, therefore, the homesteads of loan farms, after all available crown land was distributed, had to be placed at a distance that varied between one and two hours from each other, not forgetting that the nature of the soil and the distribution of the water had to support them. In this way, the principle of a minimum intervening space of an hour therefore came to work in practice against the distribution of farms nearer than one hour from one another, while at the same time it functioned as a factor that eventually prevented the distribution of farms at a greater distance than two hours from one another. As long as the Cape colony was thinly settled, however, many homesteads lay much farther than two hours from one another, while by far the majority of them were more then an hour removed from one another. According to Barrow, even at the beginning of the nineteenth century the average distance between loan farms was roughly two hours (fig. 2).[37]

The question now arises as to what distance a one hour's walk represented. In the deputies' reports distances are generally mentioned only in terms of time, without any indication of the speed with which steps were to be taken. In 1735 the Stellenbosch landdrost Louwrens and his heemraden, in their report concerning the complaints of the Pasmans against a certain Rodyn, included, however incidently, the following distances in terms of time as well as paces:

20 minutes or 2100 paces
22   "   " 2210 "
25   "   " 2800 "
36   "   " 4000 " [38]

According to these figures, roughly 110 paces are stepped off per minute. The length of a pace as well as the speed with which it is stepped off, differs, however, with different persons and it changes even with the same person in different circumstances. A calculation of distance simply in terms of paces or time was therefore not efficient. In order to obtain an objective standard, distances had to be calculated in terms of constant paces as well as time. In time, then, it became the practice to set the speed with which to stride for the stepping off of loan ground at one hundred paces of three feet or twenty-five rods of twelve feet each per minute. It is difficult to say when this practice was generally recognized, but in 1815 Charles D'Escury, the Inspector of Crown Land and Forests, declared that this was the case "from time immemorial."[39]

Farm Boundaries under the Loan Farm System

The boundaries for the grazing land that a farmer could use by virtue of his loan place lease were never indicated in this not very informative document. Apparently the government had originally never expected a license holder to confine himself within a fixed boundary. The concept "farm"—in the sense of an area with boundaries on all sides—was only connected much later with the grazing license. This is clear from the stipulation, "to be careful

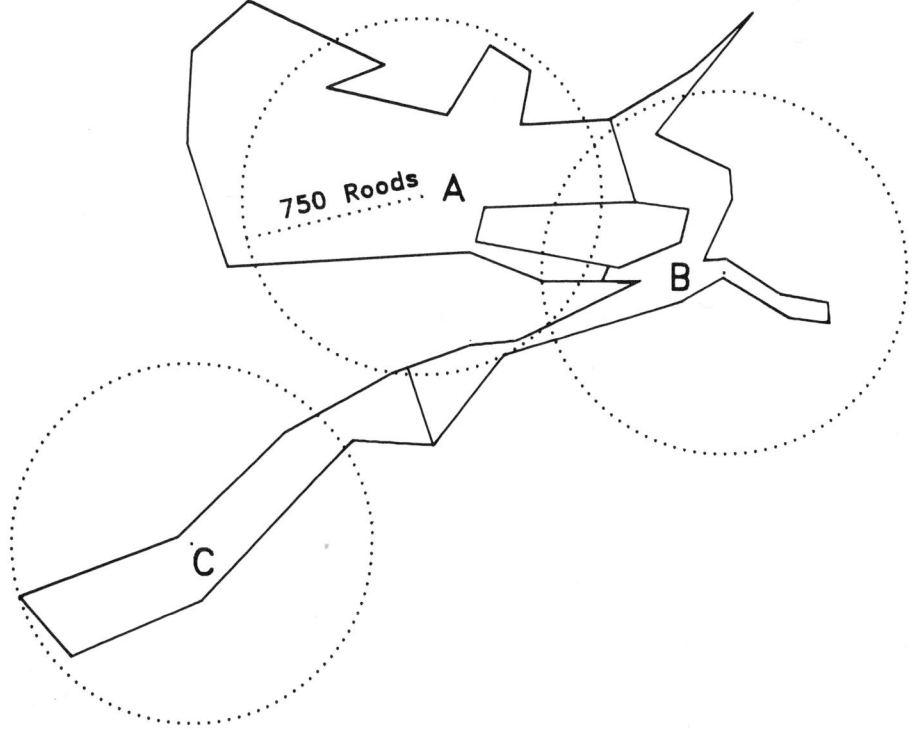

Fig. 2. These three loan farms, which lay between steep, rocky mountains in the district of Clanwilliam, were surveyed in 1831 to be distributed in perpetual quitrent.

A. Hendrik van der Watts Gat, alias Klavervallei (2,038 morgen 55 square roods).

B. Elandskloof (571 morgen 525 square roods).

C. Kafferskraal (746 morgen 334 square roods).

D. Represents a rocky mountain.

[Trans. note: A "square rood" was a standard of measurement formerly used in South Africa equal to 148.752 square feet or 14.08 square meters. 600 square roods equaled one morgen. One morgen equaled roughly 2.11654 Royal standard English acres.]

that they are not obtrusive to some already settled there," that was included on all loan place leases.

This stipulation would have been unnecessary if a *farm* was allocated by lease, with the condition that the holder may graze within certain fixed boundaries and may not graze outside of certain fixed boundaries. If a similar farm was already inhabited by someone else with permission from the government, it could on no account be allocated. And if no one else possessed it, the leaseholder could not harm anyone so long as he remained within the boundaries of his own farm.

This stipulation, which continued to appear in all loan farm leases until the end of the Company period, could naturally in the course of time have become a meaningless phrase in a stereotypical formula, but originally each section of the grazing license must have had a certain meaning. On account of the form of the loan farm lease we must therefore assume that a "loan farm" originally had no fixed boundaries. In fact, this stands to reason, since a loan farm did not have a predetermined size.

However, as other farmers were also granted grazing rights in the vicinity of a lease holder's homestead, his loan farm inevitably took on fixed boundaries as time went on. It goes without saying that as soon as two farmers, each of whom possessed exclusive grazing rights around his own homestead, went to live at an accessible distance from one another, the land in between them had to be divided into two distinct and separate grazing areas. If this did not happen there would be constant friction over pasture. And if the two homesteads lay so near to each other that the government deemed it inadvisable to allow someone else to erect a third homestead between the two, then the line that divided the pasture had to form a communal boundary for the two farms.

Thus, with the increased population density over a period of time and under the flexible system of grazing licenses, there naturally came into being—through the division of pasture between farmers—fixed "farms," the shape and size of which were determined solely by the nature of the ground and the location of other loan farms in the immediate vicinity of the established homestead.

A farmer never knew beforehand precisely what shape a fixed loan farm would eventually take. This he would only see for the first time after the government had allotted other farms by leasing in all directions around his homestead at the shortest distance that it had determined adequate. Considering, however, that the homesteads of loan farms, as a result of the nature of the land and the distribution of water, could not always be erected at the same distance from each other, it was therefore impossible that all loan farms could eventually have had the same size. Also, all loan farms whose boundaries were formed in a natural manner could not have had the same shape—first, as a result of the irregular distribution of homesteads across the surface of the land, and second, because geographical obstacles frequently prevented a farmer from grazing the same amount of pasture in all directions round about his homestead. In many cases loan farms even took on very irregular forms (fig. 3).

When the loan farm system was repealed (1813), there were already many loan farms that were bounded on all sides. They were in the older settled districts, such as the Cape and Stellenbosch, that could also support a dense population. Other loan farms had fixed boundaries only on one or two sides, where they bordered on nearby farms. In other directions, where neighbors did not live at an accessible distance, the "grazing range" was still "indefinite." Such was the case even where deputies established a boundary between two neighbors, because they generally only divided the ground between the two properties and never fixed the other borders of a loan farm. In the arid northwest, where homesteads lay far from each other and a farmer's nearest neighbor was often hours remote from him, many a loan farm proprietor did not know where his loan rights ended and unallotted crown land began. Neither the farmers nor the government bothered themselves over this impreciseness during the period of the loan farm system. Only where two loan farms bordered on one another, and differences over pasture could therefore develop, were the farmers anxious to know how the boundaries of their farms ran.

If the farmers could not reach an amicable settlement on the mutual division of pasture, help was requested from the government, whose decision then was final. No attempt was ever made, however, to fix the boundaries between loan land and unallocated crown land. It was not to the farmers' advantage to insist on this,

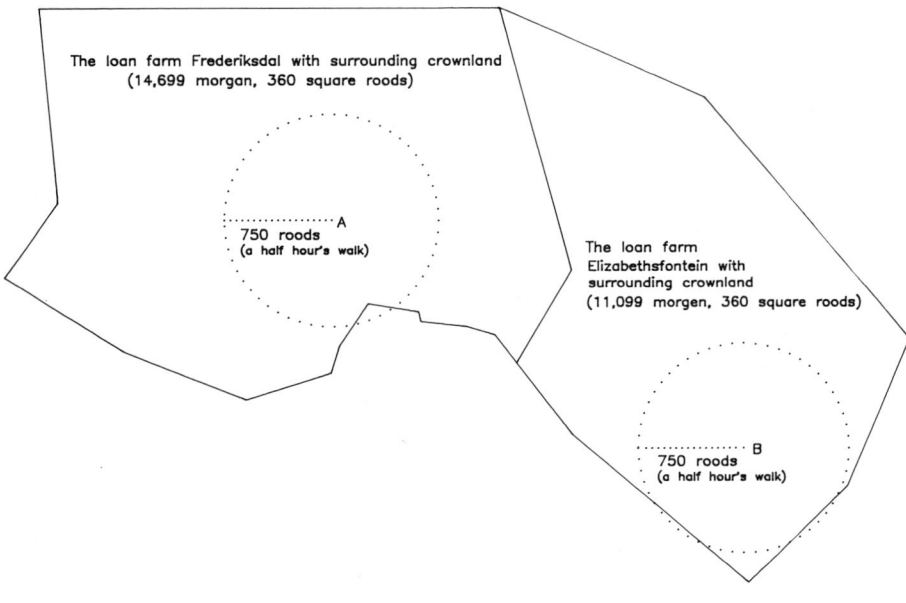

The loan farm Frederiksdal with surrounding crownland
(14,699 morgen, 360 square roods)

..........A
750 roods
(a half hour's walk)

The loan farm
Elizabethsfontein with
surrounding crownland
(11,099 morgen, 360 square roods)

..........B
750 roods
(a half hour's walk)

Fig. 3. These two loan farms in the district of Clanwilliam were surveyed in 1827 for Abraham Mouton, Sr. From the diagram it is clear that the homesteads lay more than two hours' walk from each other. The circles indicate the surface area of the land to which the leaseholder theoretically had rights in loan.

and the government was not concerned about farmers making temporary use of crown land.

Circular Farms with a Half-Hour Radius

Over time there developed among the farmers the view that a loan farm lease gave them grazing rights on a circular "farm" with the homestead at the center and a radius of a half hour's walk, about 3,000 paces, or 750 roods. A loan farm would therefore have had a size of around 2,945 morgen. It is not known precisely when this theory developed and how quickly it gained favor. Roughly a hundred years after the distribution of the first grazing licenses, mention was repeatedly made in the archival records of this theory.[40] Evidently then, it had already been generally accepted long before by the farmers and recognized by the deputies.[41]

In 1810 M. C. Gie, who had tried to determine the origin of loan farms using archival records and verbal information from old farmers, wrote to Acting Secretary Bird:

> From all leases and papers respecting to the same which have been duly examined, I can't trace why Loan Places have obtained the right of half an hour's extend each way from a certain placed Beacon. . . . This custom of giving a half hour each way has been followed for many years and serves as a limit, and now become a surreptitious custom, of which no Law exists, but from the time elapsed can be considered as a tacit acquiescence of the same, as many decisions of Landdrost and Heemraden and of Commissioners of the Court of Justice in case of disputes between two farmers have rested upon this point, and accordingly been decided in favor of one of the parties.[42]

By this time the license holder no longer complained that "the grazing and water for his livestock was harmed," but that someone else lay so many minutes from his beacon—"and therefore within his half hour." By 1827 it was not even unusual for a farmer to ask the landdrost to allow his field-cornet to measure off his "half hour!"

The Company government was not responsible for the development of the theory about circular loan farms. During the Com-

pany period there was never a fixed shape or size prescribed for
loan farms. Even when the landdrost and heemraden from Graaff-
Reinet received instructions in 1786, with reference to applications
for new loan farms, to investigate whether such a farm "can be
issued without considerable injury," no mention is made regarding
the size of a viable farm on which one could earn a livelihood.[43]
The intention of these instructions was obviously that each case—
as before—had to be investigated and dealt with on its own merits.
This was also no more than reasonable, seeing that the same yearly
quitrent was paid for good and bad farms. It could thus not be
expected that all farms must be precisely of equal size. Furthermore, it would be absurd to prescribe the same shape for all loan
farms because the natural condition of the land would not have
permitted this everywhere.

In the "Ordinance Concerning the Management of the Country Districts," wherein the Batavian government in 1805 required
a preliminary investigation for the issuance of loan farms in general,
there was also a significant silence regarding the size of a farm that
could support a farmer. This ordinance prescribed that the field-
cornet concerned had to select a fixed middle point and subsequently had to determine whether a half-hour plot of pasture could
be allotted from there in all directions. Even if this was possible,
the farm in question still could not be recorded as "sufficient" for
that reason alone. The field-cornet still had to investigate various
other points mentioned in the instructions and submit a written
report about them to the government. A farm also could not be
recorded as "insufficient" if there was *not* a half hour of pasture
available in all directions round about the middle point. In such a
case the field-cornet had to check precisely how much land there
was available in different directions and submit a written report on
his findings "to the Government and then leave the matter to them,
in order that they might rule over the sufficiency or insufficiency,
according to the circumstances."*

It appears from this that by the beginning of the nineteenth
century, the government, under the influence of well-developed
popular opinion, regarded the area comprising a circle with a radius
of a half hour's walk as more or less the natural size of a loan farm.

---

*[Trans. note: No citation is given for this paragraph in the Afrikaans edition.]

But on the other hand it is just as evident that the government did not prescribe—as per the farmers' theory—a circle with a radius of a half hour's walk as the statutory shape and size of a loan farm. The government recognized that the location of existing farms, the distribution of natural water and the nature of the pasture, made the application of geometrical principles impossible when issuing loan farms. In each particular case it wanted to take local conditions into account. For that reason the government took special precautionary measures to insure that the popular view of the farmers did not exercise too much influence on the field-cornet when he made his decision regarding the sufficiency of loan farms.

The English government followed the Batavian government's example. In 1811 the Cape fiscal* did not want to recognize the right of leaseholders to a half hour's pasture in all directions around the homestead. He conceded that the theory was generally accepted.

> But however generally this is asserted, and even confirmed by some magistrates, yet I cannot coincide therein; because there are many places, which cannot be extended to an hour in diameter without injury to other places, and because if there is no law, as is the case, prescribing this distance, there does not exist any reason, why this distance should be considered as natural; it being moreover necessary to view the business in this light, as otherwise many places must be considered as not having their legal extent, which since many years have been possessed as fully sufficient. . . . There are some possessors on loan who claim a greater distance of ground for use, than an half hour from the middle point. . . .

He then offered the suggestion that where there could not be a half hour of land in all directions from the homestead for fixed license holders, they had to be content with whatever amount of land they could get. If, however, it did not encroach on the rights of others, an established license holder must be permitted to use a half hour's land in all directions. If, however, they laid claim to

*[Trans. note: The Afrikaans edition reads "In 1811 the Chief Justice at the Cape . . ." referring to J. A. Truter. Truter became fiscal in 1809 and became chief justice on 26 August 1812. He held that office till 1828. *DSAB*, I:805. This letter is also in Theal, *Records*, VIII:91–107. Truter signs himself as "Fiscal."]

more than this, "the right thereto should appear to the satisfaction of government and in want of sufficient proof not admitted." With reference to new loan farms that would be distributed in the future, he suggested "(that) government should not by any means bind themselves to any fixed extent in the granting of lands on loan."[44]

In 1813 the question regarding the legal size of a loan farm became an issue of great practical importance. The English government decided in this year not to allot any more land in loan, and to give out in perpetual quitrent to the owners of loan farms who had applied for it as much land as they previously possessed *legally*. In the proclamation that announced the new system of land ownership, the size of a loan farm was not stated. The only stipulation was: "No loan place shall exceed three thousand morgen, every addition to that quantity of land must be particularly mentioned to the surveyor and commission, and appear on the face of the application for His Excellency's consideration."[45] This sensible stipulation, which was altogether in agreement with the above-mentioned pronouncement of the fiscal,* was implemented in such a manner that the prevailing uncertainty at the time, with reference to the lawful size of loan farms, did not in actual practice cause too many problems.

With reference to the shape of perpetual loan farms the colonial secretary gave the opinion that such land had to be surveyed in as circular a shape as possible,[46] but in answer to a memorandum from the Governor[47]—in which this letter was enclosed—the Supreme Court suggested: "that the land surveyor shall not be bound to a circular measurement, and that a circular figure shall not even be surveyed, when by that the disposition of someone else over any section of land, however small, could be subject to obstruction."[48] This letter from the chief justice to the governor was then forwarded to the various landdrosts and land surveyors for their "information and guidance."[49]

How Did the Theory of Circular Farms Originate?

From the preceding discussion it is clear that the popular view regarding the shape and size of a loan farm did not have one or

---

*[Trans. note: The Afrikaans edition reads "Chief Justice." As noted, Truter was chief justice by 1813, but not when he made the 1811 pronouncement.]

other written law or formal instruction to thank for its origin. Undoubtedly it was also not something that was formulated because it had occurred so often in actual practice that it could be regarded as an inviolable custom. A circular loan farm could only have occurred in practice if this special shape was prescribed by the government, which was definitely not the case. A circular farm could not have originated in a natural way through division of the grazing land between neighbors.

Assume, for example, a hypothetical case whereby such exceptional circumstances were present that the government could issue a farm in loan precisely one hour's distance in any given direction from an ordinance holder's homestead. Even then the government would not be able to issue more than six farms by ordinance around the homestead in question, because *their* homesteads naturally had to lie one hour from one another as well. Now, after the pasture was divided between the neighbors, the original license holder's farm would take on the shape of a regular six-sided figure and have a somewhat greater surface than a circle with a half-hour radius.

If such six-sided loan farms frequently occurred in actual practice, one could perhaps seek in this phenomenon the origin of the circle theory. Such a figure could be reshaped very easily into a circle without it causing a great reduction in area. Even though it was geometrically possible, the nature of the land and the distribution of water often made it impossible, however, to maintain the principle of an hour's intervening space when issuing so many farms around an established homestead. Besides, at the time when the theory must have developed there was also not much reason to do this. The colony was not very densely populated then and for very understandable reasons the stock farmers had never displayed any tendency to locate their homesteads as near as possible to one another. The circle theory could not, therefore, have been aligned in this way with existing practice. How then did it originate?

The theory was undoubtedly born out of the need that was felt for something more concrete and more objective than the vague and subjective concept "sufficient farm," which was so elastic that the farmers never knew precisely how far their grazing rights extended. Since the size of a loan farm was neither prescribed through regulations nor mentioned in the ordinance, however, they had to figure out the matter for themselves. They now assumed—and rightly too—that a leaseholder, by virtue of his lease, possessed grazing rights to all the land around his homestead that the government had not issued in loan to other farmers. And in this connection they had repeatedly observed that the government withdrew new loan farms that were located nearer than an hour from an old homestead if there had been complaints against the issuing of the farm. Naturally this did not always happen, but yet it was a common enough occurrence that the farmers gradually came to consider it as a fixed rule.

According to this principle of the minimum intervening space, which was actually based on observation, the farmers now developed this further through abstract reasoning. Apparently the reason a leaseholder extended his grazing rights in any particular direction at least up to a half hour's walk from his homestead was because a neighbor in any particular direction could never settle at a distance shorter than one hour from the homestead. They then erroneously concluded that a leaseholder's grazing rights extended in every direction around his homestead up to a half hour's walk, without taking into consideration that neighbors would not be able to come and live in *all* directions precisely at an hour's distance from a homestead.

Instrumental in the development of this theory was the fact that the simplest farmer, in those days when land surveyors were scarce and loan farms not surveyed, could determine precisely how far his grazing rights extended in a particular direction with minimum labor and without any instruments. In addition, a circular farm with the homestead as the middle point was very convenient for the individual farmer, because his livestock could then easily graze the entire farm without unnecessarily trampling all over the land.

On the other hand, the great drawback of such farms—namely, the practically insurmountable problems connected with surveying the land and the absolute impossibility of indicating the boundary lines of unfenced circular farms on the landscape—did not hinder the development of the theory, because during the time when the theory came into being and gradually took root, it was never actually put into practice in its fully developed form. During the period of the loan farm system the farmers were never compelled to graze their livestock only a half hour's walk from their loan farms; they never had to mark off, fence in, or survey their circular loan farms; they never tried to keep out strangers from their convex farm boundaries or to respect the concave boundary lines of their nearest neighbors.

Consequently, the theory of circular loan farms remained a hypothetical abstraction long after it was already generally accepted by the farmers. In reality the only influence it exercised was in thwarting the issuance of loan farms at a distance shorter than one hour's walk from one another.

The farmers evidently developed the theory simply and solely with the unconscious intention of giving concrete expression to the principle of an hour's intervening space, because in practice they invoked only this element of the theory constantly; on the erroneous, imaginary structure that was built on this real element, the farmers never based any practical claims. This is, in my view, the psychological explanation for the fact that a theory having so little basis in reality could originate and find general acceptance.

THE INFLUENCE OF THE LOAN FARM SYSTEM

Security of Ownership under the Loan Farm System

It was often claimed in the past that the uncertainty of ownership under the loan farm system was detrimental to the economic development of the Cape Colony—more precisely, that it stood in the way of the capitalization and intensive cultivation of the land. A farmer, went this line of thought, would never devote as much effort to a loan farm as he would to land that he had full ownership rights to, since the farm could be taken from him at any time and given to someone else. In addition, the risk connected to the investment of capital on another man's land would have made the farmers afraid to introduce permanent improvements on their loan farms that would yield profits only after a number of years, because they never knew whether their children would be able to enjoy the fruits of those efforts. Rather, they would have been more inclined to practice overcropping for the sake of immediate gains. Instead of trying to raise the value of their farms through the application of labor and capital, they would have attempted to extract all they could from them, then abandon them and request new ones. Under these conditions attachment to the land could not develop, and this was yet a further reason for overcropping and repeated change of residence.[50]

This contention was not based on testimony from loan farm owners but rather on abstract reasoning. In addition, the sharpest critics of the loan farm system were officials who were not sufficiently familiar with local conditions and were not well enough informed about the loan farm system as it worked in practice. Consequently, the question arises as to whether the claim of loan farm owners to their land was truly as uncertain as is sometimes professed, and whether the farmers truly felt that their ownership rights were precarious.

A license holder's title to the land he used originally rested exclusively on his grazing license. And according to his grazing license he received simply and solely the right "to be able to settle and graze" his livestock for a specified time at a specified place. There was nothing in the contract to which the farmer could refer in order to make a claim for renewal of the property. On the contrary. The license holder was even warned on the permit "under no circumstances to migrate." Thus, in reality, the first license holders possessed no guarantee that after their license expired they would again receive the right to go and dwell on their temporary livestock posts. Their grazing rights, therefore, were strictly temporary.

In time, however, it was inevitable that the grazing license had to lose its temporary character. Where a farmer had worked and built on a livestock post with the permission of the government, he could reasonably expect that the government would permit him to renew his license. And his claim to a license became stronger as the temporary livestock posts gradually developed into places of residence. For farmers possessing no freehold land elsewhere, it was necessary to erect their permanent dwellings on their loan farms, to build pens there for their stock, and to cultivate the ground if they wanted bread, vegetables, and fruit. The government connived at the cultivation and building on loan farms—although fully aware that they took place—and thereby left the farmers with the impression that they had official approval. It placed the government under a moral obligation to take into consideration the cultivation and improvement of the land when it had to decide whether or not a farmer's term of lease should be extended.

In reality the government never refused the regular renewal of grazing licenses for farmers whose livestock posts were permanently in use. The following case would lead one to deduce that from very early on the government welcomed this. From 1706 Jacob ten Damme possessed a livestock post in Groene Kloof, initially on the basis of six-month licenses and later on the basis of yearly permits. In 1711 his license was again renewed, but with a note attached that he had to apply for its renewal each year.[51] Herein lies not only the assumption that the license holder was going to occupy the livestock post permanently, but the note appears as well to contain the tacit promise that the extension of grazing rights would not be denied without a special reason.

In 1714 the government's direct interest in the renewal of grazing licenses was increased by the imposition of a yearly quitrent. And considering that all farmers paid equal quitrents, it no longer mattered to the government which farmer possessed grazing rights to a specific livestock post. Consequently, no objection could be raised if farmers inhabited the same livestock posts year after year. The main point was that they could not use the pasture without first having paid quitrent for it. This is clear from a 1715 ordinance, in which the colonists are warned to renew their grazing licenses at the proper time. It is furthermore worthy of note that colonists slow to renew were threatened with a fine—not with suspension of their licenses.[52] All loan farm leases thereafter stipulated that the holder thereof must renew it within a month after the expiration of the loan period. All of this together weakened the clause in the grazing license that stipulated that the farmer could not claim an extension of grazing rights.

In the oldest Gamehunters Books one often finds notes for the renewal of old leases, but in time the practice of applying for the yearly extension of grazing rights gradually fell into disuse. It was apparently regarded as self-evident that under normal circumstances the government would not deny the extension of grazing rights. Accordingly, the farmers did not bother themselves over the yearly renewal of their licenses and the government did not insist on it. Finally the practice developed whereby the loan farm leases were renewed automatically each year until either the farmers had "discarded" the farms or the government had "withdrawn" the farms. In the majority of cases it was the farmers who terminated the agreement. There were so few exceptions to this rule, that gradually the farmers must have gained the impression that the extension of their loan farm leases depended only on their own volition. Governor Janssens's 1805 memorandum regarding loan farms stated as much: "No man takes a loan farm here. He actually hires the use of a large piece of land for one year from the government, with permanent option to continue the lease in subsequent years."[53]

The government never formally waived its right to withdraw loan farms, but in practice this was done only in special cases.[54] For example, when the issuance of a new loan farm harmed an established lease holder, then the new farm was generally withdrawn. In such cases, however, the government did not make arbitrary use of its power but rightly protected the occupants of loan farms that were issued in a regular manner, and so strengthened the farmers' trust in the permanent possession of their land. Besides, in such cases the farmers in all probability would not have been aware that the government had made use of its right, as legal owner of all loan farms, but rather that the government acted by virtue of a stipulation that was included in all loan farm leases—namely, "to be careful that they are not obtrusive to some already settled there."

The government often withdrew loan farms that had been granted in the regular manner, but that never occurred without very sound reasons. It happened, for example, when the government wanted the farm concerned for the establishment of a new drostdy, or use it in some other way for the benefit of the community. In such cases, however, the lease holder suffered no harm, because the government always paid compensation. And the liberal way in which this was done could easily have created the impression that the government considered the license holder as owner of the land. Not only was the value of the homestead normally paid out, but what the entire farm was worth according to an impartial appraisement.[55] In addition, such special cases resulting in the withdrawal of loan farms generally never left the leaseholders feeling insecure about the ownership of their farms, because they were not touched by the causes that led to it. Where compensation was paid out for loan farms that were withdrawn for the benefit of the community, the government was ultimately only making use of a right that it could exercise with regard to any particular fixed property in the colony. It could not, therefore, allow the leaseholders to feel more insecure in the possession of their farms than the land owners.[56]

Loan farms were withdrawn by the government so seldom and under such reassuring circumstances that the average farmer never felt that it could happen to his farm—so long as he complied with his part of the contract and regularly paid his quitrent.

Even when the farmer fell into arrears with the payment of his quitrent, there was still no great danger that the government would take his farm and lease it to someone else. The leaseholder could keep his farm and pay his overdue quitrent later. The Company government was very lenient regarding the recovery of this money—if not extremely careless. Every five or six years the names of farmers who were in arrears with their quitrent were sent to the landdrosts, and the farmers notified that they must pay up. But the farmers generally did not bother themselves too much about this because the government had never taken legal action against them for such debts.[57]

The long periods for which the quitrent on some loan farms was in arrears is truly surprising. By chance I made notes of the case of Johannes Vosloo, Jr., who was fourteen years and three months in arrears on both his farms.[58] But there are other cases where farmers never paid quitrent for twenty or even forty years.[59] It will be clear from the following figures how common it was that farmers fell into arrears with their quitrent. In 1793 there were 1,959 occupied loan farms in the colony, and of those the amount of overdue quitrent was the substantial sum of 324,067 rix-dollars.[60] Thus, every loan farm in the colony was on average about seven years' quitrent in arrears.

If the farmers were not able to pay their quitrent as a result of causes over which they had no control—for example, Bushman robberies or one or another natural disaster—they were often exempted from it for a few years. Sometimes overdue quitrent was even canceled in such circumstances.[61] Such experience must have strengthened considerably the farmers' trust in the fairness of the government and their faith in the security of their land possession.

Yet another factor contributing a certain degree of permanence to the theoretically temporary loan farm leases was the government's recognition from the very beginning of the leaseholder's ownership of the improvements he made on his loan farm. Gradually it became an established custom that leaseholders sold their loan farms to one another or disposed of them in their wills as if they were private property. In theory only the homestead was sold and the land remained the property of the Company. But in reality possession of the land was the buyer's main goal. Consequently, the mere value of the homestead alone did not determine the price of the farm. The purchase price was generally equivalent to the value of the entire farm, including the value that the contracting parties attached to the use of the land. Thus, although ostensibly only the homestead was sold, usually the land, which was still the property of the government, was paid for as well. Often the improvements to a loan farm were not worth even as much as the transfer tax that had to be paid for the alienation of the homestead. A cottage that was worth no more than twenty or twenty-five rix-dollars was frequently sold for 20,000 to 25,000 guilders. Under the name "homestead," there were even loan farms sold on which absolutely no improvements had been made.[62]

The government was fully aware that such sales took place but did nothing to prevent them. Thus the government created the impression that it approved of such transactions. Accordingly, a buyer could reasonably expect that the government would grant to him by lease, and to no one else, the farm on which his purchased homestead stood. Indeed, it is clear that the purchase of the homestead—as well as the price paid for the farm—was based on the assumption that this should be the case, because without the farm the buildings would be of very little value to the buyer—at the most as much as the usable material he could carry away after the buildings were demolished.

But although the government could refuse to grant in loan to the buyer of a homestead the farm on which it stood, in practice— so far as known—this never happened.[63] The natural result was that in the course of time farmers came to regard the granting of grazing rights to land around a purchased homestead as something they could count on with certainty. This is obvious from the fact that they bought homesteads and paid an entire farm's purchase

price for them even before the farms were truly granted to them
in loan. A buyer was well aware that it was necessary to take out
a new lease in his own name, but because in actual practice it was
always granted, the farmers gradually came to view this as a mere
formality, without considering the possibility that the lease could
be refused. Eventually they came to believe firmly that their grazing
rights were based on their having purchased the "farm"—and not
on the lease—and that after the farm was purchased the rest was
only a matter of formal registration.

The farmers' firm conviction that the purchase of a homestead
brought with it automatic grazing rights on the farm was strengthened still further in 1790 when the government levied a transfer
tax of 2.5 percent on the purchase price of a loan farm at alienation.[64] This tax, in the farmers' eyes, did more than legalize the
sale of the homesteads. For the farmers the fact that the government
received 2.5 percent on the full purchase price of the loan farm
was a clear indication that the government, although fully aware
that this tax was paid not only for the homestead, but also for the
use of the land, had silently agreed that a loan farm owner could
transfer his right to the use of the land to someone else. The question could even be asked whether the government still retained its
right to deny the transfer of the lease in the name of the buyer
after the imposition of the tax.

In addition, the imposition of the transfer tax undoubtedly
must have strengthened considerably the farmers' trust in the security of land possession under the loan farm system. To be sure,
nothing was implied in the lease in which the tax was announced
that a loan farm purchaser, who had paid the transfer tax, had more
rights than his predecessor who had not paid the tax. But the mere
fact that the government granted official sanction to the *de facto*
transfer of grazing rights, which was based on the assumption that
the buyer of the loan farm would not be disturbed later in the
unimpeded use of his land, must have set the farmers to thinking
that they could now count on the virtual permanence of their loan
farm leases.

Indeed, it appears that by the imposition of the transfer tax
the government formally recognized the rights the farmers had
acquired through time by permanent occupation of the land. The
Inspector of Crown Lands and Forests in 1814 rightly asserted that

the 2.5 percent transfer tax for the alienation of loan farms must be regarded as "*a bonus* for allowing this transaction and the consequent occupance of the Land,"[65] but from the Resolutions of the Council of Policy it seems very clear that the government at this time did not regard the matter in that light. The Council of Policy recommended a similar tax, considering that "hitherto whenever the fixed use of loan farms was transferred from one occupant to the other, the Company enjoyed no advantage from the so-called purchase money for the homesteads of the loan farms promised or paid," whereas on the alienation of other fixed property a transfer tax was paid. In addition, it was found "that very little money would be paid for the homesteads of the loan farms when the buyer cannot assure himself that he will receive the loan farm when he requests it, because of and for which the price is actually being paid, and toward which value the purchase money was directed, even if no cottage or any framework for the homestead was found on the loan place." Therefore, the Council of Policy came to the conclusion that a transfer tax for the alienation of loan farms could be levied with the greatest fairness.[66]

From the above quotation it seems the government regarded the imposition of this particular tax as desirable because the permanent use practically did away with the difference between the sale of loan farms and other fixed property. Therefore, by the imposition of the tax the government tacitly approved these practices and placed on itself the obligation to accept the logical consequences of this action.

From the previous discussion it would appear that in time the leaseholders acquired greater claims to their loan farms than one could have inferred from the first grazing licenses. To be sure the grazing license never lost its original form, and up until the beginning of the nineteenth century it continued to form the written legal basis on which the rights of a loan farm owner were based. But by the time the government decided not to issue any more new loan farms, a leaseholder could be quite sure that the government would not hinder him in the unrestricted use of his loan farm without sound reasons and sufficient compensation.

The farmers could claim this unrestricted use on the basis of fairness. The farmers needed to cultivate and build on their land under the loan farm system and the government shut its eyes to this. It would have been very unjust if the government later denied the renewal of a specific lease without sound reason. It would have been unreasonable as well if the government had first permitted the farmers to sell their loan farms to each other and then suffer damages later when these farms were withdrawn.

Perhaps the government, by allowing the sale of loan farms, even forfeited the right to withdraw such purchased farms again without a sound reason. Undoubtedly the farmers could not transfer to each other more rights than they themselves had. But the appearance was created that these sales met with government approval because, while fully aware of the sale of grazing rights, the government never raised an objection and even received an income from them.

On the official side, it was later denied that as time went on the government had lost its right to revoke loan farms without reason.[67] But it also was publicly recognized that the loan farm owners' claim to the permanence of their leases had become so strong that no reasonable government could disregard it. If an ordinance was withdrawn, it would have jolted the leaseholders' confidence in government justice and fair dealing.[68] Indeed, the government also had such respect for these claims that not even in 1813 did it make it compulsory for old license holders to exchange their loan farms for perpetual quitrent farms.[69]

Although the farmers could not perhaps rely on the permanence of their loan farm leases on the basis of the established rules of law, the continued existence of the government's abstract right to withdraw loan farms without reason was not of any practical importance anyhow with respect to the building on and cultivation of loan farms. Indeed, it was not the purely legal position, which the farmers then understood wrongly, but their own interpretation of the matter—the position *as they understood it*—that would have influenced their behavior.

And apparently the farmers had complete confidence in the permanence of their leases. In 1805 Governor Janssens stated that the occupant of a loan farm

can consider himself, and his descendants, as peacefully settled. . . . the majority of the farmers have so little knowledge
of the difference between such a loan and a free property, that

many of them, when one speaks of it, have no understanding
of it or, if they grasp the meaning, express an eagerness to have
the loan in freehold changed, only because they imagine they
would be freed from the payment of such a quitrent, and thus
should have the land, without bearing any responsibility for any
taxes on it; because that is the way it is with the freehold land.[70]

This assertion was confirmed in 1810 by the Receiver of Land
Taxes: ". . . the possessors of loan farms, notwithstanding the annual renewal of the same is expressed in the grants, feel themselves
equally secure in the possession thereof as if they were real
property."[71]

And since the farmers did not regard themselves as vulnerable
to the eventual loss of their loan farms, the "uncertainty of ownership" under the loan farm system did not act as an inhibiting factor
in the cultivation and improvement of the land. This became clear
at a special conference held in 1803 by Governor Janssens on the
subject of cultivated land and attended by O. G. de Wet, W. S. van
Ryneveld, E. Bergh, H. Cloete, and P. A. Myburgh—all high-
ranking officials and prominent farmers. In response to the question, "Is it *advisable* or *not advisable* to make it easier for the occupiers
of loan farms to procure for themselves freehold property?" the
following decision was taken:

On the whole it is not advisable, because the government would
relinquish a right from which relinquishment neither the government nor the tenant would enjoy a benefit: *not* the government because there is scarcely a holder of a loan farm to be
found who has not obtained the same through sale or other
onerous title, and thereby paid the value, or credited to his
account; not the tenant, because no positive difference is made
by this, in the matter of work, because he possesses the loan
farm with ever so much security as a freehold property; there,
in addition to the yearly quitrent, he would have the value of
the place paid to his predecessor, and the fortieth penning purchase price paid to the state.[72]

A few years later W. S. van Ryneveld, who had traversed the
entire country as a member of the Commission for Stock Raising
and Agriculture, and therefore had had plenty of opportunity to
get at the truth, wrote:

> In fact the effect of this good faith on the part of the farmers
> is so strongly manifested, in the trouble and expenses some of
> them have been using for the improvement both in buildings
> and cultivation of the ground. In my present journey I found
> places of that description so far advanced in point of buildings
> and cultivation as any freehold place in the Colony.[73]

The feeling of security in connection with the possession of
their loan farms, which was already so strongly developed in the
colonists by the beginning of the nineteenth century, obviously did
not originate all at once. It must have evolved gradually. Precisely
when the moment was reached when this feeling was strong enough
to overcome the fear of the eventual loss of a loan farm cannot be
determined. It is, nevertheless, very evident that that degree of
security had already developed fairly early in the eighteenth century because from the outset everything was conducive to its development. The almost negligible use that the colonists made of van
Imhoff's loan freehold farm system leaves one to surmise that by
about 1743 the farmers already had as much security of possession
of their loan farms as they desired.[74] While in 1814 there were
altogether 2,291 loan farms in the Cape Colony, there were just
fifty-six entire and eight half-loan freehold farms.[75]

Thus, we can conclude that land ownership under the loan
farm system was not so uncertain as is often maintained; that the
farmers never doubted the permanence of their leases; and accordingly, that the theoretically temporary character of the leasing
rights under the loan farm system did not make the farmers afraid
to cultivate their farms intensively and to make improvements on
them by the application of capital and labor.

Other Drawbacks of the Loan Farm System

Apart from the supposed "uncertainty of possession," there were
also yet other objections brought forward against the loan farm
system. First, it was pointed out that all farms, in spite of differences in usefulness and local advantages, paid the same quitrent.[76]
This was indeed unjust because in different parts of the country
land values varied greatly. Some loan farms were best suited for
agriculture; others could only be used for livestock grazing; some,

because of the cold, scanty rainfall, or lack of a permanent water supply, could only be used a few months each year; while loan farms on the borders were continually subjected to Xhosa attacks and Bushman robberies.

It is true that the government tried to equalize these differences as far as possible. Border farmers who were greatly troubled by the natives were often exempted from quitrent payments. The government winked at the use of undistributed crown land in districts where nature obliged the farmers to migrate each year with their livestock. In some parts of the country the government even tacitly permitted the farmers to develop on their own a system of winter farms on crown land. Apparently the unformulated government policy with regard to loan farms was that every leaseholder must have enough pasture and water for his livestock. If he could not find them on his loan farm, he could go and look for them elsewhere. But this leniency on the part of government still did not entirely justify the imposition of a fixed quitrent on all loan farms.

A second drawback of the system was that loan farms did not have a uniform or fixed size. So long as the colony was still thinly populated this did not matter too much. But clearly with the growth in population density it was inevitable that such an elastic system of land ownership would give rise to innumerable disputes between loan farm owners. What is more, the convenient uncertainty of loan farm size, especially as long as there was still plenty of space between farms, continually offered the farmers an opportunity to encroach on undistributed crown land, and through the unrestricted use of this land to make permanent their grazing rights on it. The government would reap the bitter fruits of this policy when the lack of new land obliged it to grant crown land on lease between existing farms. Established leaseholders would turn up then with claims to grazing land that the government could not recognize but also could not disregard. Such a system of land ownership, which was not based on farms with fixed borders, could therefore work effectively under pioneering conditions, but the undesirability of the system would become evident as soon as conditions became more stable.[77]

The fact that loan farms could not be let out[78] was certainly an important limitation on the ownership rights of the leaseholders. At the same time, however, it had the beneficial effect of restricting the possession of loan farms to men who truly farmed, and thus protected the colony against land speculators. De Mist pointed out in this regard that the loan farm system was detrimental to public credit and money circulation in the colony. In his opinion it would have raised the farmers' credit if they had received total ownership of their loan farms, because it would be easier to borrow money against a mortgage on freehold property than against a homestead that was exposed to dilapidation and fire hazard.[79] This argument did not take into account that a loan farm had a greater market value than the mere worth of the homestead. But even if the conversion of loan farms into freehold land raised the credit of the farmers, credit was not of so much importance in the old days. At that time stock farming depended much less on credit than today.

Lastly, it was regarded as an annoying restriction that loan farms could not be subdivided.[80] This restriction had its advantages and its drawbacks. In some livestock districts fragmentation of the land would have led inevitably to impoverishment, because the carrying capacity of the soil would not have justified it. A farmer could often barely make a living on one farm and some even needed two. In agricultural areas, however, it was economically feasible in many cases to subdivide loan farms and, if this could have taken place legally, it would have given rise to more intensive land cultivation and a denser population in the colony.[81] In truth, loan farms were often *secretly* divided. Although the *lease* could only be taken out in one person's name, there was no objection raised if the homestead was owned by more than one person. Accordingly, it sometimes happened that different persons had shares in a loan farm and farmed together on it. This state of affairs was unsatisfactory, however, since the shareholders never enjoyed the full benefit of a lawful division and such a communal farm was not always conducive to the improvement of the land.[82]

The Loan Farm System and the Expansion of the Colony

It was frequently claimed in the past that the fact that loan farms could not be subdivided was an important reason for the rapid expansion of the colony. Since the family farm could not be divided following the death of the parents, each one of the children would have been obliged to take a new loan farm by lease, and in this way a wide dispersal of the population became necessary. In 1811, for example, the colonial secretary wrote to the governor,

> The population of the Colony is rapidly encreasing [sic], every new married couple seeks a spot whereupon a livelyhood can be earned in independence, but few of these are to be found except on the waste Lands for the loanlands not being the property of the holders they have at present no right to subdivide them and although on most of the loanlands a subdivision would afford undertenants a comfortable livelyhood, yet it being unsafe for them to attempt to settle thereon, a distribution of other Land becomes absolutely necessary.[83]

This reasoning was undoubtedly correct, but we must also take into consideration that loan farms were, in many cases, not economically subdividable. In actual practice, therefore, the prohibition against the subdivision of the land had less influence on colonial expansion than perhaps might appear in theory.

Of much greater importance was the fact that under the loan farm system the Company was not following a deliberate policy in regard to systematic land division. In reality, farms were granted without any formality to anyone who requested them, whenever the applicants concerned wanted them. How far a new loan farm was located from the nearest established farm was of no importance. After all, the Company was not engaged in colonization: its intentions at the Cape were purely commercial. What is more, the government did not have at its command effective administrative machinery and sufficient power to exercise effective control over the occupation of crown land by the migrant farmers. If the government refused to grant a farm by lease, the applicant would use the land anyway and the Company would be deprived of the quitrent. Accordingly, the government allowed the farmers to erect beacons for their loan farms wherever they wanted them. They had only to refrain from harming the grazing rights of established license holders.

The consequences of this policy are apparent. Considering that the accepted practice under the loan farm system was for a leaseholder to make use of half the land between himself and his neighbor,[84] a dispersal of homesteads at a greater interval from each other than the approved minimum intervening space was to the farmers' advantage. Under these circumstances the aspiring lease holder was not tempted to establish his homestead at an hour's distance from his nearest neighbor if he could just as easily lay out a farm an hour and a half from there. Naturally the farmers often based their choice of sites for their loan farm homesteads solely on the existence of a permanent source of water,[85] but this was certainly not always the case: "partly by accident, though frequently by design," Barrow reported, "the stakes are so placed that, on an average throughout the Colony, the farms are at twice the distance, and consequently contain four times the quantity of land allowed by Government."[86] Thus, the loan farm system promoted scattered settlement and for that reason the expansion into the interior occurred at a quicker pace than it would have had the government systematically divided land at the Cape according to reasonable settlement patterns.

In another way as well the loan farm system promoted migration into the interior. As some pieces of land were significantly more suitable for grazing and cultivation than others, even those located within small distances from one another, the colonists in a pioneer region initially occupied only the very best land, and left the average land between existing farms, which was not as well supplied with water, to lie unoccupied for the time being. For all farms, however, an equal amount of quitrent had to be paid. These two conditions must have inclined young newcomers in areas of older settlement to migrate across the border where good ground was still to be had, and there to seek out for themselves the best farms.

Moreover, a leaseholder on the borders had much more grazing land at his disposal than someone who had taken undistributed crown land on lease in areas of older settlement. He could not, by virtue of his lease, make legal claim to all this land, but as long as there were no other leaseholders living in his immediate vicinity, nothing prevented him from using as much grazing land as he had need of.

Lichtenstein, for example, makes mention of the case of field-

commandant Abraham de Klerk, who lived at the source of the
Gamka River. His nearest neighbor was twelve miles* from him:

> The disadvantages of living in so isolated a spot can only be
> made up for by the land's extraordinary fertility, which can
> support a flock of 8,000 sheep, and afford pasturage in the same
> degree for a proportionate number of horses and cattle. The
> main advantage consists of the owner's ability to appropriate
> land for miles around entirely for his own use, and over which
> he is absolute ruler of an area equal to many a principality. If
> the sheep have grazed one field entirely clean, their owner has
> only to drive them from there to another some distance away,
> where the grazing is abundant, and this constant change, this
> roaming about, is precisely the means by which the sheep thrive
> and prosper. With such resources, any great losses are easily
> overcome or replaced. I spoke with Commander de Klerk the
> following year in Cape Town, and he related to me, quite
> calmly, that a heavy cloudburst had caused 3,000 of his sheep
> to drown in one day. He hoped, however, to recoup the losses
> in less than two years, without suffering any particular
> privations.[87]

Perhaps Commandant de Klerk was especially lucky, but generally speaking a prospective leaseholder could rest assured that
across the border, at least for a few years, he would not have many
problems with his neighbors over pasture, and that in the immediate vicinity of the farm on which he lived he would have enough
space for local grazing and seasonal migrant grazing. The practical
value of a new loan farm was therefore greater on the border than
in the older settlement areas, not only because there was better
land available but because the farmers there did not even see the
smoke of their neighbor's chimney. This was an important incentive
that allowed the rapid dispersal of extensive stock farmers to take
place over a large area.

Finally, we can show that land under the loan farm system
was granted *so* cheaply, that it was possible for everyone who felt

---

*[Trans. note: "Twelve German miles or sixty miles English measure." H. Lich-
tenstein, *Travels in Southern Africa*, Van Riebeeck Society, series I, vol. 11, p. 41,
fn.]

like it to begin stock farming.[88] There was nothing to pay for a new loan farm and there were also no surveying costs connected with the purchase. The farmer merely had to pay the yearly quitrent, which certainly did not make excessive demands on his capital. The system facilitated a rapid expansion of stock farming, which in return gave rise to a migration into the interior.

Concluding Observations

With the passage of time a new form of land ownership, commonly known as the loan farm system, developed out of this amorphous system of grazing licenses. It cannot be determined when this system took the form that it had at its abolition in 1813 because it evolved very gradually. The development took place so unnoticed that it cannot even be determined when the loan farm system replaced the system of grazing licenses, from which it differed greatly. Perhaps we ought not even to make a distinction between the system of grazing licenses and the loan farm system, because in truth we are dealing with *one* system, which in the beginning was very flexible, but gradually took on a fixed shape.

In addition, it is interesting that the loan farm system did not owe its origins to a consciously planned set of regulations, for the entire system developed spontaneously. The grazing license maintained its original, typical form until the loan farm system was abolished. But in the meantime the contract between ruler and subject was supplemented and modified in different respects by all sorts of customs, some of which the government recognized as established practices, and various activities that met with its approval.

The first grazing licenses granted to their holders the right only to move around on common pasture with their livestock in a vaguely designated area. Under the loan farm system farmers gradually obtained exclusive grazing rights around specific points.

What is more, over time license holders received the right, by assertiveness on their part and resignation on the part of the government, to cultivate their loan farms—a right they certainly did not possess originally. Thus, as time passed, purely undefined grazing rights developed into leasing rights on specific farms.

Under the loan farm system the farmers also clearly had the right to build on their loan farms and to consider themselves as the owners of these established homesteads, although nowhere in their leases or elsewhere was this right expressly granted. They even received the right to transfer, through the sale of these homesteads, the grazing rights on their loan farms to one another. And finally, from the beginning of the eighteenth century, established practices, which could have led in time to the passing of certain legal regulations, contributed to the loan farm system, originally based on temporary and revokable concessions, evolving in the direction of a system of perpetual quitrent. Perhaps in 1813 this evolution was still not complete in legal terms, but to all intents and purposes the system was undoubtedly in place.

NOTES

1. C 1, Resolutions, 10 October 1655.

2. C 12, Resolutions, 25 February 1678, pp. 440–41.

3. C 1, Resolutions, 24 August 1678, p. 593.

4. C 3, Resolutions, 25 February 1678, p. 440.

5. Ibid., 24 August 1678, p. 593.

6. C 2271, Original Ordinance Book, 19 October 1691 [date signed], 22 January 1692 [date published], pp. 103–4.

7. C 6, Resolutions, 4 August 1687, p. 156.

8. 1/STB 19/25, Notices, 12 September 1686.

9. C 6, Resolutions, 19 October 1691, p. 401.

10. C 2271, Original Ordinance Book, 19 October 1691 [date signed], 22 January 1692 [date published], p. 104.

11. RLR 1, Old Gamehunters Books, part I, October 1687 to October 1712.

12. F. Valentyn, *Oud en Nieuw Oost Indiën* (Amsterdam, 1727), V(II):48.

13. Willem Adriaan van der Stel, "Korte Deductie," in H. C. V. Leibbrandt, *Defence of William Adriaan van der Stel*, (Cape Town, 1897), p. 12.

14. Adam Tas and van der Heiden, "Contra-Deductie" (1712), in Leibbrandt, *Defence of William Adriaan van der Stel*, p. 246.

15. RLR 1, Old Gamehunters Books, part I.

16. C 2271, Original Ordinance Book, 19 October 1691, p. 104.

17. Cf. RLR 1, Old Gamehunters Books, pp. 157, 174.

18. Ibid., pp. 141, 146, 151, 165, 173, 174, etc.

19. AR, Asiatic Council 301, report of a conference on agriculture held by Governor Jansens on 31 October 1803; M. C. Gie to Bird, 23 November 1810, in G. M. Theal, *Records of the Cape Colony, 1793–1831* (Cape Town, 1896 and 1911) VII:430; F. Le Vaillant, *New Travels into the Interior Parts of Africa in the years 1783, 1784 and 1785* (London, 1786), I:37.

20. RLR 1, Old Gamehunters Books, p. 192.

21. CJ 2963, Minute Judicial Attestation and Acts Book, Testimony of Jan Vosloo, 12 April 1708.

22. The winter-grazing farm system and the summer trek to the Agterveld will be dealt with more fully in P. J. van der Merwe, *Trek. Studies oor die Mobiliteit van die Pioniersbevolking aan die Kaap* (Cape Town, 1945).

23. S. F. N. Gie, *Geskiedenis van Suid-Afrika of Ons Verlede* (Stellenbosch, 1924), I:168.

24. C 172, Resolutions, 19 July 1786.

25. C 214, Resolutions, 4 March 1793, p. 64; C 217, Resolutions, 23 July 1793, p. 137; BR 504, Original Ordinance Book: Extract Resolutions, 29 February 1804, p. 410; BR 390, Ordinance concerning the Supervision of the Interior Districts, 23 October 1805, art. 267.

26. 1/STB 1/21, Resolutions, 3 March 1794, p. 114; C 223, Resolutions, 22 April 1794, p. 57; CO 2567, Graaff-Reinet, Stockenstrom to Governor, 28 August 1809; GH 23/2 Dispatch Book, Caledon to Castlereagh, 16 October 1809; J. Barrow, *Travels into the Interior of Southern Africa* (London, 1806), II:379. A number of interesting letters with reference to this point are found in CO 2614, Letters Rec. from Tulbagh.

27. See 1/STB 1/132, Reports from Deputy Heemrade.

28. C 547, Letters Rec., van Ryneveld (Swellendam) to Governor, 16 August 1779.

29. C 535, Letters Rec., Report from Landdrost Mentz and Heemrade, 26 November 1774, p. 597.

30. 1/STB 1/132, Reports from Deputy Heemraden, 23 March 1770.

31. 1/GR 12/1A, Field-cornets' letters, J. Potgieter to Landdrost Bresler, 5 February 1799, p. 268.

32. C 495, Letters Rec., Horak to Governor, 26 July 1756, p. 415.

33. 1/STB 1/132, Reports from Deputy Heemraden, 12 April 1776.

34. BO 49, Letters from Stellenbosch, J. J. Pienaar to Landdrost, 10 November 1796, pp. 81–84.

35. 1/STB 1/132, Reports from Deputy Heemraden, 1 August 1770.

36. C 535, Letters Rec., Report from Landdrost Mentz and Heemraden, 26 November 1774, p. 597ff.

37. Barrow, *Travels*, II:380.

38. C 1234 Petitions, reports. etc., Louwrens to Governor, 28 November 1735, new pp. 129–31.

39. CO 79, Inspector of Lands and Woods, C. D'Escury to Bird, 13 November 1815. See also the unsigned and undated document in CO 53, Surveyor of Lands, 1813; Inspector of Lands and Woods, P. S. Buissine to Bird, 9 July 1818 and C. D'Escury, Notes on the Reports upon lands in the district of Swellendam, 3 November 1826; CO 4439, Correspondence on Land Tenure (1810–1811), Extract of a Report submitted by the Landdrost of the Cape District to His Excellency the Governor, The Earl of Caledon (undated–[1810]), pp. 17–19.

40. AR, Kol. Arch. 4373, Letters and Enclosures, J. W. Janssens to G. K. van Hogendorp, 19 January 1804, p. 263; GH 23/2, Dispatch Book, Caledon to Castlereagh, 16 October 1809, p. 116; C. D'Escury, General View, [ca. 15 March 1823] in Theal, *Records*, XV:330; CO 48/13, Public Records Office, London, Henry Alexander to J. Cradock, 13 January 1812.

41. CO 4443, Fiscal to Secretary to the Government, 28 June 1811; M. C. Gie to Bird, 23 November 1810, in Theal, *Records*, VII:430.

42. M. C. Gie to Bird, 23 November 1810, in Theal, *Records*, VII:430.

43. C 172, Resolutions, 19 July 1786.

44. CO 4443, J. A. Truter (Fiscal) to Bird, 28 June 1811.

45. J. Cradock, Proclamation, 6 August 1813, in Theal, *Records*, IX:205; CO 64, Inspector of Lands and Woods, C. D'Escury to Bird, 2 August 1814.

46. CJ 2501, Court of Justice, Letters Rec., H. Alexander to J. Cradock, 25 September 1813, pp. 216–19.

47. CJ 2501, Court of Justice, Letters Rec., J. Cradock, Memorandum, 5 October 1813, pp. 220–21.

48. GR 8/3, Letters Rec. from Government, J. A. Truter to J. F. Cradock, 23 November 1813.

49. CO 4835, J. Cradock's Letterbook, Circular to Landdrosts, 14 January 1814, p. 121.

50. C 2357, Instructions from van Imhoff, 25 February 1743; AR, Asiatic Council 298: de Mist to Governor and Council, 30 May 1803; Asiatic Council 301: de Mist to Janssens, 21 December 1803; Asiatic Council 301: de Mist to Governor and Council, 10 October 1803; AR, Accession 1913 (van Hogendorp), Janssens, Memorie 30 November 1805; CO 48/13, Public Records Office, London, Truter to J. Cradock, 11 February 1812; Truter to Bird, 11 February 1812, in Theal, *Records*, VIII:269; J. Cradock to Earl Bathurst, 23 August 1813, in Theal, *Records*, IX:224;

D'Escury, General View of the Land Tenure at the Cape of Good Hope, in Theal, *Records*, XV:333.

51. C 8, Resolutions, 7 April 1711, p. 82.

52. C 9, Resolutions, 8 June 1715 and Notice of 18 June 1715.

53. PSB, Mss. Germ. Quarto 857, Janssens, Memorandum on Loan farms, 31 January 1805.

54. CO 173, Receiver of Land Revenues, List of revoked loan farms.

55. AR, Accession 1913 (van Hogendorp), Janssens, Memorandum, 30 November 1805; M. C. Gie to Bird, 23 November 1810, in Theal, *Records*, VII:431; Commissioners of Inquiry: Report II, Upon the Finances at the Cape of Good Hope, 6 September 1826, (Cape Town, 1827), p. 80.

56. C 191, Resolutions, Memorandum of O. G. de Wet, 3 March 1791; C 1813 and 1814, Letters Disp., C. J. de Graaf to Kamer Amersterdam, 19 April 1786; AR, Accession 1913 (van Hogendorp), Janssens, Memorandum, 30 November 1805; CO 2567, Letters Rec., from Graaff-Reinet, Stockenstrom to Caledon, 13 September 1809; M. C. Gie to Bird, 23 November 1810, in Theal, *Records*, VII:431; H. Alexander to J. Cradock, 13 January 1812, in Theal, *Records*, VIII:246; van Ryneveld to Bird, 24 January 1812, in Theal, *Records*, VIII:259.

57. C 1019, Enclosure, Report from G. Goetz to Governor Rhenius, 11 December 1792, pp. 393–453.

58. C 137, Resolutions, 1759, p. 144.

59. CO 173, Receiver of Land Revenues, List of revoked loan farms.

60. C 217, Resolutions, 23 July 1793, p. 32.

61. PSB, Mss. Germ. Quarto 857, Janssens, Memorandum, 30 January 1805.

62. C 171, Resolutions, 19 April 1786, p. 454; C 191, Resolutions, Memorandum from O. G. de Wet, 3 March 1791, p. 430; PSB, Mss. Germ. Quarto 857, Janssens, Memorandum on Loan farms, 30 January 1805; M. C. Gie to Bird, 23 November 1810, in Theal, *Records*, VII:431; CO 4443, Letter from Fiscal to Deputy Colonial Secretary, 28 June 1811; van Ryneveld to Bird, 24 January 1812, in Theal, *Records*, VIII:258; CO 64, Inspector of Lands and Woods, D'Escury to Bird, 22 November 1814; D'Escury, General View of Land Tenure at the Cape of Good Hope, in Theal, *Records*, XV:332.

63. Van Ryneveld to Bird, 24 January 1812 in Theal, *Records*, VIII:258.

64. C 2290, Original Ordinance Book, 20 July to 10 September 1790, p. 201.

65. CO 64, Inspector of Lands and Woods, D'Escury to Bird, 22 November 1814.

66. C 171, Resolutions, 19 April 1786, p. 451.

67. Cf. CO 4443, Truter (Fiscal) to Bird, 28 June 1811; Bird to J. Cradock, 31 December 1811, in Theal, *Records*, VIII:227.

68. J. Cradock to Alexander, 6 December 1811, in Theal, *Records*, VIII:204; J. Cradock to Earl Bathurst, 23 August 1813, in Theal, *Records*, IX:224.

69. J. Cradock, Proclamation of 6 August 1813, in Theal, *Records*, IX:204–8.

70. PSB, Mss. Germ. Quarto 857, Janssens, Memorandum on Loan farms, 30 January 1805.

71. M. C. Gie to Bird, 23 November 1810 in Theal, *Records*, VII:431. Also see Alexander to J. Cradock, 13 January 1812, in Theal, *Records*, VIII:246, and Donkin, Explanation, 29 September 1822, in Theal, *Records*, XV:86.

72. BR 444, Minutes of Commissioner-General de Mist, Account of a Conference on Agriculture, 31 October 1803, pp. 94–95; AR, Kol. Arch. 4365, Truter to de Mist, 13 November 1803.[Trans. note: That would be 2.5 percent of the purchase price. In other words, every fortieth penning would be paid to the state.]

73. Van Ryneveld to Bird, 24 January 1812, in Theal, *Records*, VIII:259.

74. A. J. H. van der Walt, *Die Ausdehnung der Kolonie am Kap der Guten Hoffnung, 1700–1779* (Berlin, 1928), p. 65.

75. CO 197, Inspector of Lands and Woods, Enclosure of "Sketch of Land Tenure in the Colony" from 1814, D'Escury to Plasket, 26 September 1825.

76. J. Cradock to Alexander, 6 December 1811, in Theal, *Records*, XIII:204; D'Escury, General View of the Land Tenure at the Cape of Good Hope, in Theal, *Records*, XV:333.

77. J. Cradock to Alexander, 6 December 1811, in Theal, *Records*, VIII:202; Dundas to Lord Hobart, 12 December 1801, in Theal, *Records*, IV:119.

78. C 199, Resolutions, 20 December 1791, pp. 663–64; C 2290, Original Ordinance Book, 28 December 1791 to 2 March 1792, VII:396.

79. AR, Kol. Arch. 4362, Asiatic Council 98, de Mist to Governor and Political Council, 30 May 1803.

80. Commissioners of Enquiry to Bathurst, 12 October 1824, in Theal, *Records*, XVIII:485.

81. AR, Kol. Arch. 4365, Truter to de Mist, 13 November 1803; AR, Accession 1913 (van Hogendorp), Janssens, Memorandum, 30 November 1805.

82. Van Ryneveld to Bird, 24 January 1812, in Theal, *Records*, VIII:263; 1/GR 16/15, Letters Disp., Stockenstrom to Bird, 25 May 1825.

83. CO 48/13, Public Record Office, London, Bird to Cradock, 13 December 1811.

84. C 552, Letters Rec., Secretary and Heemraden (Swellendam) to Governor, 1 May 1782, p. 169; 1/GR 12/1, Private Individuals, De Bruin to Woeke, 14 February 1787 and Statement of G. Booysen, 25 March 1800.

85. J. Howison, *European Colonies in Various Parts of the World* (London, 1834), I:203, 209, 343.

86. Barrow, *Travels*, II:380.

87. H. Lichtenstein, *Reisen im südlichen Africa in den Jahren 1803, 1804, 1805 und 1806* (Berlin, 1811 and 1812), II:55.

88. I will return to this point elsewhere in another context.

III

Colonial Expansion during the Eighteenth Century

The Beginning of the Stock Farmers' Migration into the Interior

Although hunters and migrant farmers in Simon van der Stel's time had already penetrated into the interior, the settlement at the Cape by the beginning of the eighteenth century was limited to the environs of Cape Town, Stellenbosch, and Drakenstein. As a result of the development of stock farming, however, the expansion inland began in all earnestness after 1700. The government now gave free rein to the stock farmers' inclination to migrate, and the new governor even reinstituted the system of grazing licenses that Simon van der Stel had opposed till the end of the seventeenth century. Under this system the farmers occupied the interior on the basis of livestock posts, which at first were used only temporarily but were later permanently inhabited. So over time there developed a new system of land possession that was perfectly suited to the needs of stock farming and the stock farmers' wanderlust.

During the first years of the eighteenth century the stock farmers migrated northward, and remained between the west coast and the mountain range running nearly parallel with it in a northwesterly direction. Until 1708 most grazing licenses were distributed in the direction of Groene Kloof, Salt River, Dassenberg, Berg River, and Twenty-four Rivers. The farmers could still not migrate eastward because the governor monopolized the land in that direction. His livestock posts lay spread out in the present district of Caledon and that blocked the route over the Hottentots Holland range for the colonists. Following the recall of Willem Adriaan van der Stel, however, the colonists took possession of his livestock posts, and with that the eastward migration began.[1]

By 1717 the most northerly livestock posts were to be found in the Verlore Valley and along the Kruis River behind the Piketberg, while the stream of emigration flowing eastward had by then almost reached the mouth of the Breede River. But still the expansion proceeded restlessly forward. By 1725 the first colonists had already settled in the Olifants River valley, and by 1732 the most northerly farms were given out near to the confluence of the Olifants and Doorn rivers, and on the Wiedouw River. Meanwhile, the eastward expansion had also advanced rapidly. The majority of farmers kept to the south of the Langeberg, and in 1727 they reached the Breede River, in 1729 the Gourits River, and in 1730 the Great Brak River.[2]

By the end of the 1720s the colonists who had migrated both to the north and to the east had pushed on between the west and south coasts and the mountains that ran more or less parallel with the coasts northward and eastward. Shortly before 1730, however, both main streams of migrating farmers ran up against serious obstacles. In the northwest was an arid wilderness region, poorly supplied with permanent sources of water. In the east were thick forests. These made the expansion between the coasts and the mountains more difficult and consequently, by 1730, the colonists began to migrate across the mountains with their livestock along the entire expansion line in search of pasture in the more elevated steppe lands. Around 1728 the migration began by way of the Olifants Kloof in the Bokkeveld Mountains and over Mostertshoek in the Witsenberg, into the Cold and the Warm Bokkeveld. At about the same time the eastern group of colonists began to migrate through Kogmans Kloof in the Langeberg. From there they migrated yet further eastward and kept between the Langeberg and the Swartberg.[3]

The Government's Attitude toward Rapid Colonial Expansion

Simon van der Stel was the last governor who wanted to hold in check the stock farmers' inclination to migrate. Willem Adriaan van der Stel and his successors concerned themselves very little with the migration into the interior. Now and then the Lords XVII had certainly warned that a mass dispersal up-country must not be permitted,[4] but the Cape government had taken the position that the rapid expansion into the interior was inevitable. In 1730, for example, they answered in response to an admonition from the Lords XVII: there is a vast expanse of grazing land needed for the livestock, without which the agriculturist cannot manage, since he needs his draft oxen as well as manure in order to work his land. A shortage of permanent sources of open water contributes still further to the dispersal of the colonists. At a distance of thirty to forty miles* around the Cape there are no fountains or usable pools of water that are not used either on the basis of freehold property or loan farms. Disputes and lawsuits between neighbors over water are an everyday occurrence. The farmers are obliged, therefore, to travel deep into the country to search for grazing land and water for their livestock. These people are subjected to great inconvenience, and since grain cannot be transported to the Cape over untrodden paths for such long distances, the loan farms in the interior can only be used for livestock farming. But, they explained finally, it is also a fact that "this expansion of the inhabitants with their livestock is, however, a principal reason why for some time meat could be supplied so inexpensively to the Company and private individuals."[5]

Nevertheless, Baron van Imhoff, the commissioner who visited the Cape in 1743, spoke out against the great territorial expansion that was occurring at the Cape. The most distant farmers were by then already living a good hundred and twenty hours from the fort, while perhaps half the occupied land was not being used. Van Imhoff appreciated the difficulties facing a densely populated agricultural settlement at the Cape, but he still thought the expansion had proceeded further than was necessary. He denounced the rapid migration into the interior because, in his opinion, it promoted the moral degeneration of the colonists.[6]

The Cape government also recognized that it really was necessary to gain a little more control over the migrant farmers, some of whom were already beyond Mossel Bay. On van Imhoff's recom-

---

*Dutch miles [Trans. note: one Dutch mile = approx. four English miles.]

mendation it was decided to set up "a sort of magistracy" in the outlying districts, and to place a Company official there "to keep a watchful eye on the activities of the residents and to look after judicial concerns."[7] Thus in 1743 there was established in "the far outlying districts" a subdrostdy that became an independent drostdy in 1745 and three years later received the name Swell-en-Damme.

Shortly after the establishment of the new subdrostdy, the eastern boundary of the outlying districts was fixed at the Brak River, and a hesitant attempt was also made to put an end to the stock farmers' migration away from the Cape. The governor notified the deputy landdrost of Swellendam to order the persons that had migrated to the Gamtoos River with their livestock to abandon their farms and return to the colony. The deputy landdrost was at his wit's end because the migrant farmers were altogether beyond his reach,

> wherefore he submitted to the heemraden whether they might make a suggestion as to the manner in which he should best be able to notify such persons. After some deliberation they were unable to offer other suggestions, except that it might be possible to notify one or the other elephant hunter that had to pass this way; but that no other persons could be commanded to return to the district because of their remoteness, as their dwellings are 20 days' ride from here, and most of the time no homesteads or people are encountered along the way.[8]

Thus, the most distant migrant farmers escaped from government control altogether, and the government official at Swellendam, without any means of enforcing the government's instructions, could do little to prevent the further expansion of the colony.

The Eastward Migration

In the years following van Imhoff's visit to the Cape, more and more farmers migrated over the mountains, and the principal expansion of the colony now took place in an easterly direction. Even the stream of emigration originally flowing northward turned to the east during the 1740s. Out of the Bokkeveld the colonists migrated through the Karoo, which separated this region from the Roggeveld, and they settled along the northern slopes of the Roggeveld Mountains. The colonization of this region was limited, however, to the valleys and mountainous sections, where most of the permanent water sources were found. The area to the north was used as temporary pasture, and livestock posts were erected as far as the Rhenoster River and the Small and Great Riet rivers. But even by the end of the century the area where farmers had set up permanent residences was restricted to a narrow strip on and along the mountains.[9]

From the Roggeveld Mountains on, this second eastward stream of emigration went into the Nieuweveld Mountains, which were reached by 1760. Along this mountain range the expansion then proceeded further eastward. In 1770 the Sneeuberg were reached, after a few colonists had already migrated into the Camdeboo in 1769. From here the migrant farmers spread southeastward. And when some of them migrated into the land between the Sundays and the Boesmans rivers, they met the first Swellendammers there. The latter migrants led the first stream of emigration to the east, originally proceeding between the Langeberg and the south coast. After migrating across the Langeberg, they spread into the Little Karoo. This temporarily brought the eastward migration to a halt, but it began again in 1755, and by 1760* was in full swing. In 1765 the migrant farmers reached the Gamtoos River, and before 1770 the first of them arrived in the Camdeboo. In this manner the two migrating streams met in the late 1760s, having moved eastward in parallel directions from the middle of the eighteenth century.[10]

Although the expansion into the interior after 1750 mainly flowed eastward in two parallel streams, the stream that originally flowed to the north still trickled through a little in that direction. The farm Akerendam—present-day Calvinia—was issued by ordinance in 1750. After that other colonists settled round about the Hantamsberg. In addition, a small group of farmers migrated to the Copper Mountains, across the Olifants River. In 1760 there were even a few farms granted as far north as Kubiskou. Other

---

*[Trans. note: The Afrikaans edition incorrectly gives this date as 1750.]

farmers migrated in a northwesterly direction. In 1760 the Kamiesberg were reached, and a small number of farms were granted by ordinance in the surrounding area.[11] After this, however, northward expansion was halted temporarily. Only in the nineteenth century would it begin again.

Cloppenburg's Proposals

After the establishment of Swellendam, the Cape government worried little about the stock farmers' rapid spread into the interior. The first Company official after van Imhoff to again give serious attention to the matter was J. W. Cloppenburg, who was initially independent fiscal and in 1766 was appointed deputy governor. He undertook a journey throughout the colony in 1768 and was alarmed by what he saw. In the first place, he found that the colony had expanded so far already,

> that, speaking with all due respect to the Lord Governor, neither His Excellency, nor any of the Council, nor landdrost, nor land surveyor have any personal knowledge of this expansion and these remote homesteads. I must add here that either through the ignorance or the carelessness of others, the Lord Governor has been misled about the distribution of farms, for otherwise this action, which can have so many evil consequences, is not explainable or excusable in a gentlemen of such merit and discernment, of whom few people born in a century are his equal in integrity and the affairs of government.[12]

This "far extension" of the land, according to Cloppenburg, was responsible for there "clearly being a great degeneration of religious faith, of obedience, and as a result, of good behavior among the present generation," something that threatened posterity with full-scale barbarism. At Swellendam, for example, he talked with eleven men at the military exercises. From those discussions it appeared that, with the exception of one man who belonged to the Lutheran Church, not one of them, or even the wives of the nine that were married, was a member of the Church.[13]

In Outeniqualand he had an intimate conversation with an H. van der Watt over the general lifestyle of the other farmers living in that region, and his report confirmed my opinion, that many households are very slovenly, of which the very worse are those of Ehren Kroon, and Hans Dietlof. The first one keeps house with a bastard, and the second with a Hottentot. The worse of all is that of Fredrik Zeelen, a really terrible brood, with two brothers-in-law named Smith, who advised me that they had not been baptized.[14]

Later our informant received the message that Frederik Zeelen "had had his wife baptized and was married to her."[15]

In these circumstances Cloppenburg now asked himself the question, "If I found then such folk on a journey, which did not constitute half of the circumference of our possessions, how many more of that rabble shall there be farther away? What shall become of them? Troublesome inhabitants, good-for-nothing and dangerous to society." And in order to combat the evils that the dispersal into the interior had already created and to prevent greater evils in the future, he now suggested various measures that were quite impractical. The principal one was that absolutely no more farms would be distributed. This meant that poor people would be obliged, as in Europe, to go and work for the rich. (Later Cloppenburg formulated this proposal somewhat differently: there must not be any more new farms issued deeper in the country; only between the already distributed farms must new land be granted on lease.[16] Furthermore, the high government in Batavia had to be requested not to transport any more slaves to the Cape. That would allow the development of a white working class. Finally, all means had to be employed to keep the Khoikhoi out of the farmers' service. That would be accomplished principally by making the Khoikhoi on the Company's posts so comfortable that they would not be eager to go and live with the farmers(!). In this way the farmers' sons and daughters would be taught in due course to work, and the necessity for the rapid expansion of the colony would decrease.)[17]

Cloppenburg was a voice crying in the wilderness. Whether he truly exercised any influence on the government is extremely doubtful. It is coincidentally true that shortly after his journey the Cape government fixed the eastern border and tried indecisively to maintain it for a few years. But the motive for this action was not a concern for the fate of the migrant farmers. This step was taken primarily to prevent the illegal livestock exchange with the Xhosa, which it was feared could lead to hostilities with that nation.[18]

The Influence of the Khoikhoi on Expansion

Not only had the government done nothing during the first seventy years of the eighteenth century to keep the migrant farmers in the colony, but *before that time* there was nothing that stemmed their migration either. Only during the 1770s, when the colonists on the eastern border clashed with the Bantu and the difficulties with the Bushmen on the northern border began on a large scale, did the native peoples of the country begin to exercise a negative influence on white expansion into the interior. Until then the colonists came into contact chiefly with the Khoikhoi, and they were never a counter to white expansion. Rather, they made possible the migration into the interior and even promoted it to a certain extent.

We have already seen that the Khoikhoi, who were often troublesome in the colony's early years, were no longer a political threat to the colony by the end of the seventeenth century.[19] The smallpox epidemic of 1713 weakened their power still further. Since the whites sustained heavy losses during this time, it must have wreaked even greater havoc among the Khoikhoi, who did not live so hygienically. Valentyn reports in this regard that the Khoikhoi died by the hundreds

> so that they lay everywhere along the roads as if massacred, cursing the Dutch, who they said had bewitched them, and fleeing into the interior with their stock folds, huts, and livestock in hopes of finding a sanctuary there from the evil sickness. As a result, very few Hottentots were afterwards to be seen here (as I myself found in 1714) compared with previously.[20]

After the pox epidemic many Khoikhoi who escaped from it migrated into the interior in order to save their livestock from the unusually virulent livestock sicknesses prevalent in the colony at the same time.[21] This greatly reduced the Khoikhoi population and weakened their political might. In 1717 the Stellenbosch landdrost duly wrote to the governor that they no longer had to be afraid of the Khoikhoi and that therefore it was no longer necessary to maintain soldiers at the Company's outposts.[22] Roughly ten years later there were no more Khoikhoi settlements to be found within about 250 to 300 miles of the Cape, although here and there an isolated family or two still lived together.[23]

In the course of the eighteenth century the Khoikhoi gradually lost their economic independence, and coupled with this their political organization vanished. Through wars among themselves and also as a result of the continuous barter with the colonists, they gradually lost all their livestock and were assimilated into the economic system of the whites. By the seventies one rarely came across an independent Khoikhoi settlement, even in the outlying districts. The overwhelming majority of the Khoikhoi by this time already lived among the colonists.[24] In 1778 Governor van Plettenberg, who evidently had a certain interest in ethnology, experienced to his disappointment that even in the northeastern border districts the detribalization of the Khoikhoi was already an accomplished fact. They were already totally dependent economically on the farmers and were even forgetting all their original manners and customs.[25]

Under these circumstances it is easy to understand why the weakened Khoikhoi *could* offer no organized opposition to the white expansion into the interior. Besides, the Khoikhoi were relatively peaceful. It frequently happened that a Khoikhoi domestic servant or a vagrant slaughtered a sheep or was guilty of breaking into a house, but such a thing was of purely local importance and could be punished by the farmer himself or by the landdrost. The Khoikhoi never made the border districts unsafe by thievery or murder or drove the farmers from their farms so that a military force had to be sent out against them, as was repeatedly the case with the Xhosa and the Bushmen.

According to Mentzel, instead of resisting white expansion, the Khoikhoi instead welcomed the white presence in the interior. The farmers' firearms offered them protection against wild animals as well as against their common enemy, the Bushmen, who were the enemy of everyone that owned livestock.[26] In addition, the Khoikhoi frequently possessed little and sometimes lived in lamentable poverty; they quickly became fond of tobacco and brandy, and were consequently very happy to be in the farmers' service. For limited payment, which usually consisted only of food, tobacco, clothes, skins, and perhaps now and then a cow, they watched over the farmers' livestock by day and at night kept the fires burning around the stock pens in order to keep away wild animals.[27]

The Khoikhoi presence in the interior undoubtedly contributed much to making the migration into the country possible. This lower economic class could make a living in the interior relatively easily by working for the stock farmers. Apart from that, the Khoikhoi also offered the colonists who had penetrated into the wilderness a certain measure of protection. For that reason a young farmer usually established himself gladly in the neighborhood of a Khoikhoi settlement.[28]

Wrote Mentzel,

> The Hottentots are like bloodhounds, who hunt out the most fertile lands. When their kraals are discovered, one soon finds several Europeans or Afrikaners who, wanting themselves to settle there, through cajolery and gifts, easily obtain the Hottentots' permission to settle among them. In the course of time, however, when these grazing lands have become too scanty to support the cattle of both the Hottentots and their guests, the Hottentots are induced through small gifts to withdraw farther inland with their cattle, hunting yet further possibilities in those regions to abide.[29]

When a young farmer's living necessities ran out or he had to travel into Cape Town for other reasons, he usually took his wife and children with him out of fear that perhaps during his absence they would be murdered by the Bushmen or Xhosa. "There remains then house and hearth, cattle and all else standing under God's hand that he owns," and the farmer did not know whether his livestock would all be stolen and his farm destroyed while he was away. But generally he could trust his Khoikhoi, "and to be sure, if the Hottentots were not by nature such honest, faithful, and benign men, no European or Afrikaner could live in these regions."[30]

Mentzel exaggerates a little and perhaps sees the Khoikhoi a little too idealistically, but without a doubt they did much to make stock farming possible in the wilderness and facilitated the farmers' migration inland.

The Establishment of the Eastern Border in 1770

Until nearly the end of the 1760s the colonists of the Stellenbosch and Swellendam districts migrated in two parallel streams to the east. The Stellenboschers followed the Roggeveld and Nieuweveld mountains while the Swellendammers moved on between the Langeberg and the Swartberg. Between the two groups of colonists the Great Karoo and the Goup lay uninhabited. However, when the migrant farmers from the two respective districts traveled beyond the point where the Goup ended, they began to mix with one another. With the passage of time the people from Stellenbosch and those from Swellendam were living among one another and thus uncertainty and many times even disputes developed between the landdrosts of the different districts over the jurisdiction which each one possessed. This matter was finally discussed before the Council of Policy on 14 November 1769 and they saw the need to end this irregular state of affairs. It was consequently decided to send the two landdrosts, each one accompanied by his secretary and two heemraden from his district, to fix a precise boundary line between the two districts. The commission was also directed incidentally to investigate whether, among the recently issued farms, "there might also be some, the occupation of which should be considered inadvisable because of their too remote location."[31]

At the beginning of 1770 this commission submitted its written report. The report dealt mainly with the question of the boundary line, but there appeared

> among other matters, to our particular annoyance, that the said Commission, on their journey from the Fish River to the Gamtoos River, encountered divers persons, who possessed no farms or posts in loan from the Honorable Company in this region, grazing at will considerable herds of cattle; while others, showing no respect, wandered about hither and thither with their cattle, many days' journey from their loan farms.[32]

Furthermore, the commission found piles of manure and the remains of straw huts at different sites, which proved that the farmers had maintained livestock posts there.[33]

The Council of Policy then concluded that because of this the Company had not only lost income from loan farms, "but that such selfish activities were practiced undoubtedly with the chief purpose of making it more convenient for them to carry on an illicit trade in cattle, whether with the Hottentots living in the vicinity, or with the so-called Caffers."[34] For this the said council, among other things, found proof in the fact that the commission from the Swellendam district had uncovered a well-beaten wagon path into Xhosaland—and this notwithstanding the fact that the livestock trade with the Xhosa was forbidden on pain of physical and even capital punishment.

In order, therefore, "to root out the said evil once and for all in the land," the Council of Policy then decided no longer to grant loan farms across the Gamtoos River.[35] It was fully realized, however, that this decision still would not keep the migrant farmers west of the Gamtoos. Although the government had refused to grant loan farms to the east of the Gamtoos, the farmers would still migrate of their own accord across the river anyway. For that reason it was found necessary also to make known by means of an ordinance that no one would be allowed to settle to the east of the Gamtoos River, and the Swellendam landdrost was directed to get those who had transgressed the prohibition to turn back.[36] Moreover, the landdrosts of both border districts were notified to watch with extreme attentiveness that no one graze his livestock anyplace else than on the farms that he received in loan from the Company, "much less wander hither and thither with them, or on any other pretext, to leave their homesteads and proceed into the interior." In such a case all the livestock held in this manner would be confiscated for the Company's benefit.[37]

This last regulation was chiefly intended to make it more difficult to evade the ordinance and prevent the loss of the Company's quitrents. But the establishment of the eastern border for Swellendam was linked primarily to the government's Xhosa policy. The Gamtoos River was fixed as the eastern border and the colonists were forbidden to settle themselves beyond it, not so much because the government regarded the expansion of the colony as undesirable but because it wanted to prevent unlawful livestock trade with the Xhosa. This appears very clearly from the following: from the report of the above-mentioned commission, the Council of Policy concluded that the illegal livestock barter with the Xhosa occurred only in the Swellendam district. For that reason an eastern boundary was established only for this district, although the first Stellenboschers were already in Camdeboo, which lay much further eastwards than the Gamtoos River.[38]

The migration of the Stellenboschers into Camdeboo caused no uneasiness to the government. Further expansion of the Stellenbosch district was even considered desirable. This is clear from the fact that the Council of Policy accepted the following recommendation of the above-mentioned commission without objection:

> as there is no route (from Camdeboo) leading into Kafferland by which an illegal trade in cattle can be carried on, and, on the other hand, as nothing other than wild and savage Bushmen and Hottentots, who possess no cattle, and who must subsist solely on wild game, are living in those regions of the country, so there appears to be no evil to be feared from this side either. The undersigned are therefore of the opinion, that not only should the loan farms already present there remain in loan, but even, if it pleases Your Excellency, still more farms might be distributed, along the Bushman Mountains in the east as far as a certain Heights, which lies between the Bleije River, which is the last or most easterly branch of the Sundays River and the first branch of the Fish River, called by us DeBruijns Heights. It is true it would be better if the farms already present in Camdeboo could remain just as they are now, but since the surrounding countryside is rich in grass and productive, it can be understood why the inhabitants of these farms would take advantage of the opportunity and establish farms there, from which the Honorable Company would not receive the least income. But if, as has been stated already, farms were distributed as far as the already mentioned Heights, then the Honorable Company would receive from them the fixed loan rent.[39]

This quotation beautifully characterizes the position of the Cape government toward the expansion of the colony during the eighteenth century. In order not to lose any quitrent the government was always ready to follow the farmers on their migration into the interior. The government never concerned itself much about the expansion of the colony as such. When the government in 1770 forbade the migration of the Swellendammers over the Gamtoos River it was only to prevent the illegal livestock barter with the Xhosa, because of the difficulties that could develop from such contact with this nation.

This commercial mentality of the government, which always placed the Company's interests first, as well as the government's powerlessness to stop the expansion into the interior, is clearly evident from a letter that Governor van de Graaff addressed to the Company officials in 1786:

> . . . rigid ordinances, issued in the beginning, where possible, to prevent and to obstruct the harmful expansion of the Colony, have not been enforceable and an emigration of the inhabitants, with their wives and children, takes place every day into the wide, far distant country. Therefore the government, partly in order not to do damage to the Honorable Company's laws, and partly in order to regulate the places there and not to contribute to the inhabitants going to battle with each other over the places which they add there, and which they have gone to occupy, must proceed to distribute in loan, according to the loan rules, the places there, even those that lie nearby, to the profit of the Honorable Company.[40]

The Farmers' Migration into Agter-Bruintjieshoogte

The Council of Policy did not establish an eastern border for the Stellenbosch district in 1770, because the commission that had to go and decide on the boundary line between the two outlying districts was under the impression that the Stellenboschers *did not* take part in the forbidden livestock trade. The Council of Policy, however, quickly found out that the commission had been mistaken. This then gave rise to the desire to prevent Stellenboschers from migrating further eastward. Accordingly, the ordinance against the livestock trade, issued on 16 June 1774, specified that no one would be allowed to settle beyond Bruintjieshoogte, while the Swellendammers were warned for the second time not to migrate across the Gamtoos River.[41]

The ink on the ordinance was hardly dry, however, before the Stellenbosch landdrost had to recall a number of farmers who had settled on the other side of Bruintjieshoogte.[42] Instead of turning back they addressed a petition to the governor and remained where they were while waiting for an answer. In a pathetic manner they explained that they had learned with great sadness of the complaints that had reached the governor, that they were said to have been "obstinate and rebellious" against the government's ordinance, as well as that they were conducting a trade with the Xhosa, and that this had roused His Excellency's anger against them. They assured the governor, however, that they were obliged by their extreme poverty to migrate to Agter-Bruintjieshoogte. Many of them owned barely one hundred sheep and fifty cattle. In the region where they now lived the grazing land was good, the ground well suited for agriculture, and there was an abundance of wild game. They had not had problems with the natives. If they had done wrong to migrate into this region, they asked for forgiveness, but at the same time they asked that the governor allow them to remain where they were and to grant them farms in loan. They further promised to remain law-abiding citizens and faithful subjects, and take care that there would never again be complaints about their conduct. Finally, they requested the governor to appoint a field commandant among them, who could settle disputes and charge those who carried on trade with the Xhosa, so that they could be punished as an example to others.(!)[43]

The Land Hunger of the Swellendammers

In the meantime there was also agitation in Swellendam in favor of territorial expansion. The farmers firmly believed that their wealth and welfare were inseparably connected with unrestricted expansion into the interior, and for that reason they did not want fixed borders to restrict their movement. Consequently there was great discontent in 1770 when the Gamtoos River was fixed as the boundary, and the "precise fixing" of the district made the future there look gloomier than normal. It was just at this time that new administrative buildings were needed for the district, but it was impossible to find the necessary funds. The district was already burdened with heavy debts and high interest rates, "and in addition it is greatly to be feared, that by the spoiling of the pasture, which will greatly lessen the income of the inhabitants, as well as by the narrow fixing of the district's borders, this Colony [Swellendam] can no longer expect the revenue as has been necessary up to now."[44]

In the years that followed, farmers' complaints about land shortages increased, and finally in 1775 a special combined meeting of landdrosts, heemraden, and military officers was convened to discuss the question of the colony's expansion. At this meeting it was decided to complain to the governor about the district's needs. A detailed petition was composed in which it was shown

> how for some years past the inhabitants of this colony have submitted the following complaints: that because of the narrow limits of this district, which reach no further than the Gamtoos River, they were not able to properly expand outward with their families, which were growing larger each day. They were enclosed as well by the colonists from Stellenbosch, who have for grazing land not only the Camdeboo and Bushmanland as far as Bruintjieshoogte, which lies a few day's journey further than the Gamtoos River, but also land deeper into the interior where no boundaries have been fixed. They therefore petition to have the same freedom to migrate inland from that side.

Under these circumstances, the petition continued, the compilers of this document want to bring to the attention of the Governor and Council of Policy:

> That unless this colony expands farther east- and northward, the inhabitants will not be able to obtain for themselves or their children any more farms and therefore will not only remain in their present poverty-stricken state but must fear that it shall become worse. Because, unless the inhabitants can obtain more pasture than is now available within the narrow confines of their district, they will suffer a great shock to their welfare. And the colony's revenues, which are few and of little significance, instead of growing larger will be still further diminished so that one day they will reach a state where they will be unable to maintain the necessary buildings in proper repair, much less to settle the heavy debts with which the colony is burdened.

Because of these conditions they therefore kindly requested that the governor, "permit a small expansion" in the interests of the prosperity and welfare of the district. The proposal they produced made it clear, however, that their idea of a "small expansion" was fairly elastic. From this it is also obvious that the farmers' opinion regarding colonization did not take into account the population density or systematic land division, which would be of great importance for the internal communications and sound economic development of the colony. They suggested that

> as the spacious fields of the Camdeboo have been taken over and inhabited already, which, might we be allowed to say! with Your Excellency's and Honorable Council's gracious permission, by right ought to belong to this colony. And as the land [please note!] along the coast, even as far as the Bushmans River—would be of little or no use, despite its great expanse, because it is most generally useless, barren country, experiencing great numbers of droughts and containing large tracts of unusable forests, that would at most offer up no more than twenty farms (!); so we, the undersigned, once more humbly take the liberty of requesting that Your Excellency give consideration as to whether it might not be judged to serve in the favor of the Honorable Company together with the peace and welfare of this colony, if the lands behind the Bruintjieshoogte, along the mountains as far as the Fish River, most all being good, useful, grazing lands with rich grass, were to be added immediately to this colony in order to have a fixed boundary line there for the first time.[45]

The Expansion of the Eastern Border in 1775

When the governor received the petition of the farmers from Agter-Bruintjieshoogte, he turned for advice to the landdrost and heemraden of Stellenbosch. He sent the petition to the said council and asked them to inquire into the matter and

> consider and report on not merely whether the request in the petition can be acceded to without prejudice to the boundary line between the Stellenbosch and the Swellendam districts, but as well in what ways the daily growing population could

best be taken care of through the provision of good farms and grazing lands, which are no longer to be found in the neighboring countryside. Furthermore, they are to plan such smooth changes when altering the said boundary line, which should be considered necessary for the wellbeing of the petitioners and other inhabitants in general, as well as to bring the most benefit and profit to the two districts.[46]

This question was discussed then in the heemraad's meeting on 23 January 1775. Present at this meeting was Field-Commandant G. R. Opperman as well as the former heemraad members, Martin Melk and Jan Bernard Hoffman, who were members in 1770 of the commission to determine the boundary line between Stellenbosch and Swellendam. As a result of the discussion that took place at this meeting, the landdrost drew up a written report in which was advocated the expansion of Stellenbosch as far as the Fish River, and the expansion of Swellendam up to the Swartkops River or at the utmost as far as the Bushmans River. There was also an unclear boundary line between the districts mentioned.[47]

A few months later the governor received yet another report, which also convinced him of the necessity for the colony's boundaries to be expanded. At the beginning of 1775 Heemraad P. Myburg went to visit his livestock farms in Camdeboo and in the vicinity of Bruintjieshoogte, and the landdrost asked him to make use of the opportunity to investigate whether the Stellenboschers were obeying the governor's order not to settle beyond Bruintjieshoogte. On 1 May 1775 Myburg drew up his report. He had come across people from Stellenbosch on and nearby the Sneeuberg, and behind Bruintjieshoogte he met a group of nine farmers from Swellendam who were settled there, "and who testified that they were absolutely compelled to come here from time to time, due to the shortage of good farms on which they can find enough grazing and water for their cattle to make a living." (The majority of these farmers had signed the petition that was sent to the governor on 10 November 1774 from Agter-Bruintjieshoogte.)

In addition, Myburg suggested the following:

The undersigned, according to Your Excellency's wishes, has inquired whether our inhabitants might be in need of more farms because of the increase in families and persons in our magistracy as well as in Swellendam, and therefore whether an extension of the boundary lines of both districts ought to be made. He has not only found such an action necessary, because the aforementioned men possess barely a single good farm, but primarily, because some among them must get along and provide for themselves by living together in abject poverty on one farm, among whom he found, Willem Prinslo, Sr, Claas Prinslo, Jr, Hendrik Krieger living on one, and Johannes Kloppers, Hendrik Kloppers and Jan Nortje the elder, living on another.[48]

With these two reports and the petitions from Agter-Bruintjieshoogte and Swellendam before him, the governor had to decide whether he should maintain the clearly defined eastern border. He found that that would not be desirable and decided to expand Stellenbosch as far as the Fish River and Swellendam up to the Bushmans River.[49] There was, however, still no agreement over the boundary line between the two districts. Consequently, a combined meeting of landdrosts, heemraden, and former heemraden of both districts was convened in November to discuss the matter. The dividing line between the two districts was finally determined to be the Swartberg and an extension of the line beyond Bruintjieshoogte and along the Rietberg as far as the Fish River. The area north of that line would fall under Stellenbosch, and the area south of it under Swellendam.[50]

Van Plettenberg's Border Delimitations with the Xhosa

During the 1770s the eastward migration became of primary importance in the history of the colony's expansion, while the northward expansion was closed off altogether. In the northwest the climate and soil conditions of the country prevented the farmers from migrating farther. Along the eastern part of the northern boundary the Bushmen made further expansion impossible. They had even driven the farmers of the Sneeuberg from their lands and thereby accelerated the migration to the east.[51] During this period, therefore, as many Stellenboschers as Swellendammers migrated to the east, and consequently advanced the rapid occupation of the fertile grazing lands to the west of the Fish River.[52]

When the white expansion reached the banks of the Fish River, however, a serious situation developed on the eastern border. Here the whites came into contact with the Bantu, who were stronger and more civilized than the other natives with whom they had thus far been acquainted, and who were just as interested in the land as themselves. In 1778, when Governor van Plettenberg made his famous journey through the colony, a clash between white and Bantu already threatened. The Bantu had expanded westwards and by the end of the 1770s crossed the Fish River. Van Plettenberg realized that conflicts could develop over pasture as a result of this movement, and in order to remove this possibility he prescribed a policy of territorial segregation with regard to white and Bantu. The new eastern border was not formally proclaimed before 1786, but by 1778 van Plettenberg had already concluded treaties with different Xhosa paramount chiefs to recognize the Fish River as the boundary between the colony and Xhosaland.[53]

The Xhosa did not fulfill their side of the treaties. In spite of their promises to depart, they remained to the west of the Fish River and, by the end of 1770s, the first open clash occurred between white and Bantu. This clash ended in favor of the colonists, and did not discourage the farmers from migrating farther to the east. A few years later some farmers even abandoned their farms in the colony and fled into Xhosaland to escape from Bushmen robberies,[54] while other colonists requested farms on the other side of the Fish River.[55] Nevertheless, the first Frontier War made the government recognize ever more clearly the necessity of maintaining the prescribed segregation policy. The Xhosa could become a dangerous enemy of the colony, and war between the frontier farmers and the Xhosa would, in any case, create difficulties for the government. In the interests of the peace and safety of the colony it was best to restrict the colonists to the west of the Fish River and to forbid further expansion to the east.

The Establishment of Graaff-Reinet

By the middle of the 1780s the farmers in the northeastern corner of the colony were very troubled by Bushman robberies. By this time the Bushmen had already driven many farmers from their farms, and they threatened to chase away those that remained as well. To strengthen these pioneers in their struggle against the enemy, it became desirable to place a landdrost in their midst. This was not, however, the only reason why a new drostdy was established in the Camdeboo in 1786. A much more important motive for this step was to

> insure that no more of the remaining inhabitants from the afore-
> mentioned lands proceed into Xhosaland, and consequently de-
> prive the Honorable Company of the considerable income from
> the quitrent that their loan places provide; in order as well not
> to arrive at a point where the Xhosa drive out the inhabitants,
> and hostilities, which are carried on against them in a barbarous
> manner and have only recently come to an end, begin again
> with this populous yet peaceable nation.[56]

For other reasons too, it had become desirable to gain more control over the migrant farmers. After the Swellendam drostdy was established, the colony once more underwent enormous territorial expansion, and the first migrant farmers now lived entirely beyond the landdrost's reach. It was, in political terms, an anomalous state of affairs,

> and above all to sooner or later tempt one or another sea power,
> with which our republic might be at war, to cast their eye on
> Algoa Bay, from where an orderly government could readily
> entice destitute inhabitants in their interests. With that we sud-
> denly find ourselves with an established and blooming colony
> there that both deprives the Honorable Company of its income,
> and cuts off the supply of slaughter cattle hither. All these
> reasons taken together, make it extremely necessary that a mag-
> istracy be established in the Camdeboo, for that is where the
> best location is, even in the same way as at Stellenbosch and
> Swellendam.[57]

All the new district's borders were marked in 1786 except its northern one. On the western side the district would be bounded by the Klein-Leeuw River from its source in the Nieuweveld Mountains to where it flowed into the Gamka. From here it passed through the Gamka as far as the Swartberg, and from there along the northern side of this mountain up to the Brak River. The area embraced by these boundaries previously came under the jurisdiction of Stellenbosch. The region across the Gamtoos River, which earlier fell under the control of Swellendam, would now come under the jurisdiction of the new district as well, and be separated from Swellendam by the Gamtoos. That section of the colony bordering on Xhosaland would therefore fall under the new district, and the border between the colony and Xhosaland was also now formally proclaimed. It would run along the Tarka and Baviaans rivers to where they flowed into the Great Fish River, and from there follow the latter river to the sea.[58]

By a similar fixing of the district boundaries the eastern border problem could be placed under the administration of a single government council. The colonists were forbidden to settle on the other side of the eastern boundary, "to settle down or there to graze with their cattle on pain of immediate confiscation of all the cattle that are found there," and the landdrost from Graaff-Reinet was given orders to notify those who transgressed the prohibition, "to remove themselves from there with all due haste and to settle within the interior fixing of the boundary line."[59]

Whites beyond the Fish River

The government's segregation policy with regard to white and Bantu failed, first as a result of Xhosa activities, but also because the government failed to keep the farmers west of the Fish River. Hunters penetrated into Xhosaland and frequently hunting went hand in hand with livestock trading. And seeing that barter with the Xhosa was forbidden, it again led to farmers remaining with their stock on the other side of the Fish River. Although it was relatively safe to exchange livestock on the borders with the Xhosa, a man could still get into trouble. Now and then deputies traveled around the colony, and occasionally drove out the Xhosa cattle from among a farmer's herd and took these with them back to the drostdy.[60] Consequently, it was safer to keep bartered livestock in Xhosaland. The following example nicely demonstrates this connection between hunting, the livestock trade, and roaming with livestock across the Fish River.

In 1778 Piet, a Khoikhoi who had grown up and worked on the farm of Claas Prinsloo, explained that he, together with his master's son, two burghers, and three boys rode "far inland" in order to shoot elephants. First they went northward to the "Great River," on the other side of which they had expected to find elephants. The river was full, however, and they then decided to move southward into Xhosaland. Here each one of them exchanged livestock, "giving for each head of cattle four bunches of beads as well as two copper plates and for each calf one string of beads." After they had bartered for enough livestock they departed, but still remained on the other side of the Fish River, "on a broad plain, where no one could see them," and there they divided the livestock and went their separate ways. The storyteller and his young master left their cattle with Willem Prinsloo Senior (evidently a family member) and after that they went home. After the last plowing season, however, they returned again, took the cattle and with that crossed the Fish River. From there Piet's young master rode back to the house, leaving him behind with the livestock and, "because the lions were very daring there, a rifle, a powder horn, and some shot with which to set up a spring gun." One morning, however, when he was still asleep, Heemraad Johan Bernhard Hoffman came on his location, took away his weapon, and brought him, the other Khoikhoi who were there, and the Xhosa cattle to the drostdy.[61]

Drought and locusts also contributed from time to time to the colonists migrating over the Fish River with their livestock. In 1786, for example, the colony was visited by an "unbelievable multitude of locusts, which quickly decimated the land in various districts of this colony, whereupon then very many livestock perished from hunger and want."[62] In the same year there was a severe drought: "according to reports from some places, as in the Camdeboo, Agter-Bruintjieshoogte and the Great Fish River, there had never been a year to equal this for the drought, and also a large number of cattle and nearly all the large game there were dead."[63] On his journey through the border districts Landdrost Woeke everywhere saw skeletons lying in the fields. Under these conditions a number of farmers migrated across the Fish River with their livestock.[64]

There was a great drought again in 1792 and again there were a few farmers who sought escape across the Fish River. Others continually pestered the landdrost to give them permission to migrate as well.[65] The landdrost flatly refused their petitions and made the governor aware of his actions. The governor and the Council of Policy approved van Baalen's action and notified him,

> that the said river has been established as the boundary between our people and the Xhosa and that he keep away from all quarrels and disagreements with them. Not only must he constantly decline and reject these petitions without any hesitation, but also when finding that some persons, under the aforementioned pretext of drought, actually have proceeded over there already and have settled down, shall immediately have drawn up the necessary orders that these people promptly leave the Xhosa Territory and return to this side of the river. They will be able thereby to carefully avoid giving any reason for discontentment that might be the cause of new hostilities with the aforementioned otherwise very peaceful nation.[66]

By this time, however, it was not only fear of the Xhosa that caused the government to consistently refuse to allow further eastward expansion. Among high-ranking officials there was a growing uneasiness over the lot of the natives. Gradually they even placed nearly all the blame for the disturbances with the Xhosa on the shoulders of the farmers, who were suppose to have robbed the "peaceable natives" of their property and driven them from the land "which Providence had intended for them and the peace and possession of which is solemnly ensured by this government." In order, thus, to save the colony from the "righteous" revenge of the Xhosa and for the sake of the "preservation of the rights of man"— as declared in the language of the French Revolution—the farmers had to be kept out of Xhosaland![67]

Notwithstanding government ordinances, migrants still occasionally crossed over the Fish River.[68] They evidently visited Xhosaland only intermittently in the beginning, but some of them later settled there. In 1794 a list was sent to the governor of twenty-eight farmers who had settled over time beyond the Fish River. Among the names on the list was former Heemraad Jacobus Tregard who, in 1786, served as a deputy in announcing the ordinance regarding migration across the Fish River.[69] This migration was not of great significance, however, in terms of colonial expansion. Until well into the nineteenth century the government and the Bantu together were successful in preventing white expansion beyond the Fish River.

Northward Expansion

When eastward expansion was halted by clashes with the Bantu toward the end of the 1770s, there developed a need for expansion to the north. It was not long before the farmers on the eastern border began to complain that they could no longer acquire good farms[70] and wanted to migrate northward.[71] Little came from this expansion to the north, however, until the end of the eighteenth century.

Nature determined the boundaries in the northwesterly half of the colony for the migrant farmers. During the eighteenth century the desert prevented them from penetrating further to the north. The rainfall was too scanty and too changeable and moreover, permanent sources of open water were lacking. The farmers did in fact use the region far to the north as pasture periodically,[72] but, during the last four decades of the eighteenth century, very few farmers had settled further north than the point that had already been reached by 1760. Until well into the nineteenth century the "Agterveld" or "Grasveld" still did not have large numbers of inhabitants.[73] It was used only by the farmers of the Hantam and the Onder-Bokkeveld as communal summer pasture after thunderstorms fell there.

The country north of the Roggeveld, and the fringe of the Nieuweveld, were also too dry and too changeable during the eighteenth century to entice permanent residents. After thundershowers the region between the Riet and Sak rivers, and also the area farther to the north, could be used as pasture, but there were few farms to be found that could be occupied right through the year. Springs were scarce; the rains fell here one year and there the next; and the waterholes in the Sak River became brackish during droughts. Besides, the farmers opposed the issuing of farms to the north because they wanted to use the land in this direction as communal pasture. Accordingly, the land north of the Riet River also lay unoccupied until well into the nineteenth century.[74] The permanent colonization of the northwest had to wait for better equipment in order to dig wells, and for the waterdrills and wind-mills of our modern age.

Nature, however, left open a path for white expansion to the north during the eighteenth century, since the rainfall of the Cape Colony increased progressively from west to east. Between the de-sert and semidesert areas in the northwest and the strong and well-organized Bantu societies in the east, there lay a relatively narrow strip of land that could be inhabited without difficulty. Along this route, which ran over the Sneeuberg, down along the Seekoei River, into the present Orange Free State and Transvaal, the farmers could migrate to the north when the Bantu made further expansion to the east impossible. This, however, the *Bushmen* did not allow. Dur-ing the last three decades of the eighteenth century the Bushmen along this section of the northern boundary continually stole the farmers' livestock, murdered their herders, and frequently burned their houses. The farmers had to fight to keep already occupied land inhabitable and did not always even succeed at that. Fre-quently the Bushmen drove them temporarily from their farms. And although the farmers always returned again, there was no talk at the end of the eighteenth century of further expansion to the north. Only at the beginning of the nineteenth century did the Bushmen opposition to the expansion fade away and the northward movement of the farmers, which had already taken on such enor-mous proportions even before the Great Trek, commence.[75]

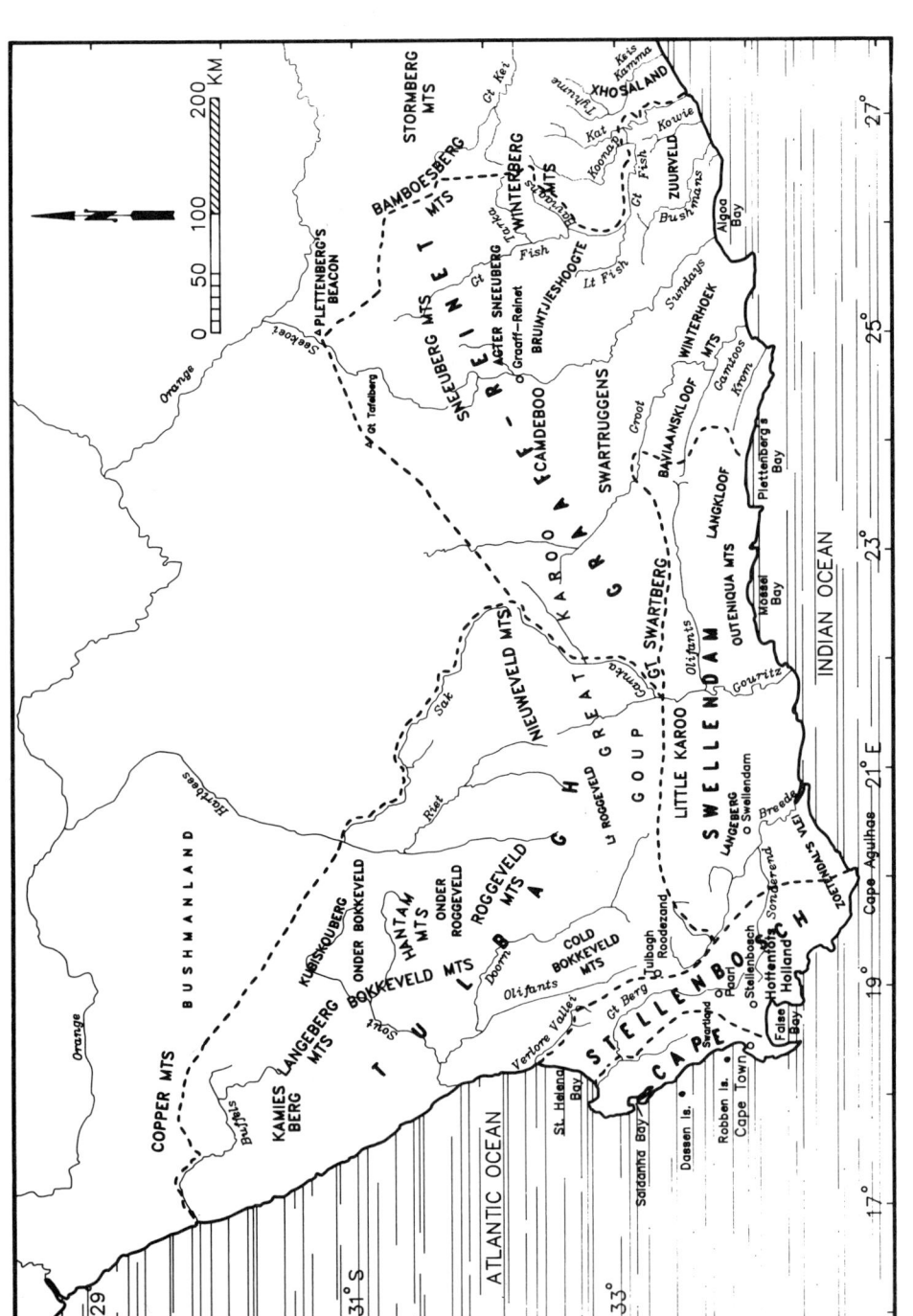

NOTES

1. A. J. H. van der Walt, *Die Ausdehnung der Kolonie der Guten Hoffung, 1700–1779* (Berlin, 1928), p. 64.
2. Ibid., p. 65.
3. Ibid.
4. C 431, Letters Rec., Lords XVII to Cape Government, 18 August 1729, p. 555.
5. C 1483, Letters Disp., Cape Government to Lords XVII, 21 January 1730, p. 103; cf. 1/SWM 1/1, Minutes, 5 January 1757, p. 202ff.
6. C 2357, Instructions from van Imhoff, 1743.
7. C 121, Resolutions, 12 November 1743.
8. 1/STB 19/165, Overberg Affairs: Minutes of heemraad meeting, Swellendam, 27 October 1744.
9. Van der Walt, *Ausdehnung*, p. 69.
10. Ibid.
11. Ibid.
12. AR, Accession 242 (Nederburg), J. W. Cloppenburg, Annotations and Remarks (1769), p. 13.
13. Ibid., p. 12.
14. Ibid., p. 15.
15. Ibid.
16. Ibid., p. 89.
17. Ibid., p. 15.
18. See pp. 118–19.
19. See pp. 23–24.
20. F. Valentyn, *Oud and Nieuw Oost Indiën* (Amsterdam, 1827), V(2):51.
21. O. F. Mentzel, *Vollständige und Zuverläszige . . . Beschreibung . . . des Guten Hoffnung* (Glogau, 1785, 1787), II:37.
22. 1/STB 20/1, Letters Disp., 3 August 1717.
23. C 1469, Letters Disp., Cape Government to Lords XVII, 25 February 1726, new p. 246. [Trans. note: The only reference in this volume for this date gives the distance as "50 to 60 miles," not "250 to 300 miles" as cited by vdM. Perhaps vdM translated Dutch miles into English miles (1 Dutch mile = roughly 4 English miles) without telling the reader.]
24. C 520, Letters Rec., Mentz to Governor, 22 February 1769, p. 485; cf. F. Masson, *An Account of Three Journeys from Cape Town into the Southern Parts of Africa* (London: *Philosophical Transactions of the Royal Society,* 1776), 66:275.

25. Journal of van Plettenberg's Journey, 1778, in G. M. Theal, *Belangrijke Historische Dokumenten over Zuid-Afrika* (Cape Town, 1896, 1911), I:11.

26. A. Sparrman, *Reize naar de Kaap de Goede Hoop, de Landen van den Zuid Pool, en rondom de Waereld* (Leiden, 1786), I:188.

27. Mentzel, *Beschreibung*, I:37.

28. Ibid., II:170.

29. Ibid., I:10.

30. Ibid., II:176.

31. C 147, Resolutions, 14 November 1769, pp. 296–97.

32. C 2285, Original Ordinance Book, 13 February 1770 [date signed], 26 April 1770 [date published], p. 381–82.

33. 1/STB 20/2, Letters Rec., Miscellaneous Report of 7 February 1770; C 148, Resolutions, 13 February 1770, p. 50.

34. C 148, Resolutions, 13 February 1770, p. 53.

35. Ibid., p. 54.

36. C 2285, Original Ordinance Book, 26 April 1770, p. 383.

37. Ibid., p. 384.

38. As he crossed over the Gamtoos River in [10 December] 1773, Thunberg noted that "at this time it formed the boundries of the Colony, and which was not suffered to extend farther. This was strictly prohibited in order that the Colonists might not be induced to wage war with the courageous and intreped caffers or the Company suffer any damage by that means." See K. P. Thunberg, *Travels in Europe, Africa and Asia performed within the years 1770 and 1779* (London, 1793), II:79.

39. 1/STB 20/2, Letters Rec., Miscellaneous Report of 7 February 1770.

40. C 1813 and 1814, Letters Disp., Van der Graaff to Kamer Amsterdam, 19 April 1786, pp. 523–24.

41. C 2285, Original Ordinance Book (part V), 16 June 1774, p. 474.

42. C 153, Resolutions, 11 July 1775, p. 284.

43. C 1264, Petitions, Reports, etc., A. H. Krugel and twelve others to Governor van Plettenberg, 10 November 1774, pp. 443–45.

44. C 2227, Logbook (Minutes of Heemraad meeting, Swellendam), 15 June 1771, p. 65.

45. C 1265, Petitions, Reports, etc., Landdrost, Heemraden and Military Officers from Swellendam to Governor, 17 March 1775. [Trans. note: The Afrikaans edition reads "om niet eenen" but a check of the original document shows that it is "om met eenen."]

46. C 153, Resolutions, 11 July 1775, p. 293; C 1265, Petitions, Reports, etc., Landdrost and Heemraden from Stellenbosch to Governor, 30 January 1775.

47. C 1265, Petitions, Reports, etc., Landdrost and Heemraden from Stellenbosch to Governor, 30 January 1775.

48. C 1265, Petitions, Reports, etc., P. Myburg to M. A. Bergh, 1 May 1775.

49. 1/STB 10/4, Letters Rec., Governor to Landdrost, 11 July 1775; C 2211, Logbook, 11 July 1775, p. 494; 1/STB 1/17, Resolutions, 7 August 1775, p. 361.

50. 1/STB 1/18, Resolutions, 13 November 1775, pp. 532–33.

51. P. J. van der Merwe, *Die Noordwaartse Beweging van die Boere voor die Groot Trek (1770–1842)* (The Hague, 1937), p. 12ff.

52. Cf. E. C. Godée Molsbergen, *Reizen in Zuid-Afrika* (The Hague, 1932), IV:15, Account of Swellengrebel's journey, 1776.

53. C 158, Resolutions, 14 November 1780, p. 358; Journal of van Plettenberg's Journey, 1778, in Theal, *Belangrijke Historische Dokumenten*, I:23, 26.

54. C 169, Resolutions, 28 August 1785, p. 599.

55. 1/SWM 10/2, Logbook, 25 October 1785.

56. C 169, Resolutions, 28 August 1785, pp. 601–2.

57. Ibid.

58. C 2288, Original Ordinance Book (part V), 19 July 1786, p. 485.

59. C 172, Resolutions, 19 July 1786, p. 874; C 2288, Original Ordinance Book (part V), 19 July 1786, pp. 482–85.

60. Cf. 1/SWM 3/14, Judicial Testimonies, Testimony of Ruiter, son of J. Erasmus, 17 January 1778; Testimony of Adam, bastard son of J. Potgieter, 17 January 1778.

61. 1/SWM 3/14, Judicial Testimonies, Testimony of Piet, 17 January 1778.

62. 1/SWM 10/2, Logbook, 17 June 1786.

63. C 563, Letters Rec., Woeke to Governor, 6 November 1786.

64. Ibid., p. 684.

65. 1/GR 589, Resolutions, 7 May 1792, p. 5.

66. C 213, Resolutions, 12 February 1793, p. 782–83.

67. C 213, Resolutions, 12 February 1793, p. 783.

68. 1/GR 3/16, Judicial Testimonies, Testimony of P. M. Bester, 28 January 1791.

69. C 223, Resolutions, 4 May 1794, p. 183.

70. C 567, Letters Rec., Landdrost to Governor, 17 September 1787, p. 473.

71. 1/STB 20/4, Letters Disp., Landdrost and Heemraden to Governor, 19 April 1785.

72. Cf. Wm. Paterson, *A Narrative of Four Journeys into the Country*

of the Hottentots and Caffraria in the Years 1777, 1778, 1779 (London, 1779), pp. 58, 60.

73. 1/WOC 12/18, Letters Rec. from Officials, van Ryneveld to Trappes, 3 July 1826; CO 2696, Letters Rec. from Worcester, G. Nieu-wenhoudt (18 December 1826), H. C. Niewenhoudt (2 December 1826) and J. Cloete (2 December 1826) to the Secretary at Clanwilliam.

74. Van der Merwe, *Noordwaartse Beweging*, p. 6.

75. Van der Merwe, *Noordwaartse Beweging*.

IV

The Tradition of Independent Farming

In 1784 a number of prominent Cape colonists produced a petition in which they summarized all the economic wisdom of their time. They complained about limited market opportunities as well as a weak and changeable market, and declared that although the colony's population was constantly growing, the availability of means of livelihood for the colonists had not increased at an equal rate. Thus, they concluded "that there was real poverty among a large portion of the colonists, who for this reason were migrating over the mountains in considerable numbers in order to maintain themselves there by stock farming, as they could not make a living as agriculturists."[1]

The colonists stress here an important reason for the rapid tempo of colonial expansion during the eighteenth and nineteenth centuries. Their conclusion, however, was based on the assumptions that the son of each farmer had to become an independent farmer as well and that whoever could not make it as an agriculturist must earn his livelihood as a stock farmer. In point of fact, during the period under discussion agriculture could not offer an *independent* existence to all colonists. But agriculture could absorb a number of white laborers, and other means of livelihood were available in older sections of the colony as well, if the colonists were only prepared to give up their prized independence and go to work for someone else. Economic factors are clearly evident in the history of white dispersal into the South African interior. No one can deny the enormous role they played. On the other hand, it would be wrong to regard the migrant farmers in the interior as the victims of inexorable circumstances over which they had altogether no control. Psychological factors certainly played a role as well in our pioneer history.

The Labor Issue in the Cape Colony

From the earliest days of the colony, agricultural areas were continually faced with labor problems. Farm workers were always quite scarce and expensive. Initially there was little use made of Khoikhoi labor. The Khoikhoi were not disposed to do the heavier sort of farm work and, what is more, they were generally unwilling to hire themselves out for it.[2] Likewise, native labor did not play a significant role in the agricultural districts before the Great Trek. Consequently, the government was compelled to begin the importation of slaves. This did not prove a satisfactory solution to the labor problem, however. Slave labor was, in general, very expensive. Large prices had to be paid for the slaves and costs were high for food, clothing, health care, and the erection and maintenance of slave quarters. In addition, slaves had to be continually imported and purchased since there were so few females slaves in the colony, and the proportion of deaths to births was about ten to three.[3]

The question now arises as to whether there were any opportunities for white workers at the Cape. If the government had never imported slaves, nothing would have prevented whites from going to work as agricultural laborers. The Cape climate was agreeable enough, and the Khoikhoi would not have offered strong competition. Furthermore, slave labor was so expensive that whites could definitely compete with it. If one studies the living conditions under which numerous stock farmers maintained an independent existence in the interior, then one cannot help asking whether they would not have been materially and socially better off had they lived in more densely populated and civilized areas and worked as common laborers.[4] In many cases people who kept independent but small stock farms in the border districts, certainly would have raised their living standards had they been willing to go and work in agricultural areas.

Even stock farming had a certain need for white workers. To be sure, looking after livestock was not the proper work for a white.

Khoikhoi and Bushmen could be obtained so cheaply that whites, who had to maintain a higher standard of living, could not successfully compete with them. But there were many farmers in the older settlements who possessed stock farms in the interior and had need of overseers. Whites were considered for these jobs because of their greater competence and trustworthiness. Sons of colonists, however, were not willing to work as foremen for other farmers. This only occurred when these sons found themselves in very bad circumstances,[5] and usually only in order to enable them later to begin independent farming on their own. Even soldiers who were given their freedom so they might work as servants, generally never remained long as agricultural workers. For the most part they received their wages in livestock and as soon as they had enough with which to begin they commenced farming on their own. Some tried to enlarge their herds by marrying farmers' daughters so they might begin independent stock farming that much sooner.[6]

This unwillingness of whites to serve as workers meant that labor requirements at the Cape could never be satisfied during the pioneer period. Population growth and white immigration did not increase the labor market supply but they certainly did raise the demand, because as soon as whites became independent farmers they had need of workers themselves.

The Alleged Laziness of the Farmers

Travelers who did not always understand the circumstances under which colonists lived frequently accused them of laziness. The origin of this alleged disinclination toward bodily exertion may be attributed to the fact that the farmers gradually got used to having slaves and natives to work for them.[7] Because the farmers' sons were too lazy to work regularly and hard, as this line of thought was further developed by others, they chose a wandering existence in the wilderness in search of stock farming rather than hiring themselves out to someone else.[8] In a private letter to Swellengrebel an anonymous writer set out this attitude very clearly:

> I also see quite well how greatly expanded the colony is. Yes, even already too far expanded and I expect that in time land shall be scarce. But why are people here not in service to their

fellow citizens, as in Europe? Why don't the young wags go
and serve as farmers' servants; while those that want, should
they not be needed, can go and live like cattle with cattle in the
fields. But let them soon understand, I believe that men, even
with at least four right hands, would have to do a lot of work
before there was much progress because, as Your Excellency
knows, the majority here have gorged abundantly from the fat
of the land, finding it more comfortable to sleep the entire day,
hanging over the bottom half of the door, waiting to see their
neighbors coming home in the evenings, than to go behind
the plow.[9]

It is very risky to make such allegations. The passing traveler often
expressed a superficial judgment about the idleness of the Afrikaner
housewife who sat at the traditional table and drank tea, while
slaves and Khoikhoi servants did the housework.[10] If he had seen
the same woman in time of danger or during moments of busy
activity on the farm, he would have altered his opinion.[11] Someone
noticing a farmer sitting listlessly on the veranda and smoking his
pipe after the livestock were cared for, would perhaps get the impression the man was lazy and lacking in energy. But if he could
have seen the same man again later trailing an antelope or stolen
livestock, he would be surprised by the liveliness and stamina displayed by the same man.[12]

Where there were specific incentives present to perform manual labor, the colonists did it without hesitation. If our pioneers
did not always lead such active lives, then this can be attributed
just as well to the fact that the stimulus to greater effort and steadier
work was lacking. Indeed, in the old days poor marketing opportunities frequently worked to undermine the initiative and industry
of the farmers.[13] As a rule, where enough slaves or natives were
available to execute specific sorts of work, the farmers mainly entrusted it to them. But this need not necessarily be a sign of laziness. What otherwise hardworking businessman will shine his own
shoes if there is a servant to do it for him?[14]

But on the other side, let us openly acknowledge it, the living
conditions of the farmers during the pioneer period were not always
conducive to the formation of diligent habits. Especially on a stock
farm there was not always a lot of work and, moreover, the regular

work was usually done by Khoikhoi or Bushmen. In 1805 the Commission for Stock Raising and Agriculture reported:

> The stockfarmer or his household knows hardly anything of the so-called pastoral life. The children occupy themselves with the cattle only to the extent of counting them now and then, or to attend to the lambs in the evenings and the mornings; or if there is danger from Bushmen or cattle stolen, to make for properly arming the place.[15]

There were certainly times of extraordinary pressure on a stock farm—for example, when the ewes had their lambs or the livestock were freezing to death in cold and rain—but for the farmers' sons there was more often than not no necessity to work hard and steadily right through the year. If the land was fertile and the Bushmen caused no trouble, then there were many opportunities to hunt and to visit.[16]

The fact that the farmers' sons frequently lacked the habit of working regularly and hard definitely hampered their ability to adapt to the life of an agricultural worker.[17] In many cases they chose to undertake stock farming because the work was not so demanding. On the frontier the pioneers had to endure numerous privations and they were exposed to all sorts of dangers, but the life was free and untrammeled. In 1805 the Commission for Stock Raising and Agriculture reported:

> The young people all appear most inclined toward stock farming, that is, as they have called it, *to migrate to the country*. This they do with 50 to 100 sheep, with which they head toward one of their family in the country, or to a place which they have here or there taken on lease. The Commission ran across this same tendency as well even in districts where children could obtain a more than comfortable existence on their parents' farms by planting rice, tobacco, corn, and such on those parts of the farms which still lay wholly untouched. The pastoral life is certainly the easiest, and to this can be attributed as well the aforementioned inclination that appears among many young

people who have no refinement or industry, and know nothing but a brutish existence.[18]

The Possibilities for an Independent Existence

It is particularly important to note that right through the pioneer period every able-bodied young man *could* find an independent existence in stockfarming. Throughout the eighteenth and nineteenth centuries this profession was open to expansion, and placed no high demands on the beginner's capital resources. The livestock necessary to begin an independent farm could easily be accumulated. A young man not possessing livestock himself could begin as a servant on a stock farm in the interior. He usually received his wages in livestock and he generally had the right to keep his own small farm. In addition, there was no need for him to butcher his own livestock, for his master supplied him with food. This enabled him to accumulate enough livestock within a reasonably short time to go off and farm on his own.[19] If a person did not want to work as a servant, it was generally not too difficult to arrange with a wealthy farmer or through a relative to acquire a few stock animals in return for half the offspring. The young farmer then had to care for the livestock and later give back the same number he received, but each year he could mark for himself half of the lambs that were born. In this manner his own herd grew larger and larger and within a few years he had a prosperous farm.[20]

The majority of farmers' sons who grew up with their parents usually already owned some livestock by the time they were old enough to get out on their own. Frequently they were stock farmers from birth, because the future of a son in South Africa was very quickly becoming associated with stock farming. By the time of Adam Tas the baptismal gift sometimes took the form of heifers,[21] and the parents supplied their children early with livestock as well. The easiest and most natural manner then to provide for a child's future was to see that he had some livestock with which to farm when he was an adult. So the tradition developed to mark for each child at birth a female calf or two and several she-lambs. This was regarded as the child's property and with the offspring could form the nucleus of a future farm. Generally by the time a farmer's son left his parent's home, he was already a well-to-do stock farmer. A marriage normally enlarged his herds substantially, since farmers' daughters usually owned livestock as well. In the interior a few stock animals could still be traded from the natives. In most cases, therefore, there was no need to purchase livestock when a farmer's son wanted to begin farming independently.[22]

The Land Policy of the Company Government

Company land policy, which was not based on sound principles of colonization, contributed still further to making possible an independent existence as a farmer for anyone who desired to do so. The Company was a commercial organization that had little interest in systematic colonization. As long as enough grain and meat could be produced, the government showed little concern about the occupation of new ground or the wanderlust of its subjects. Agricultural land was given out free of charge in order to encourage the production of grain. Stock farmers generally did not receive land in full ownership, but under the loan farm system they could get as much pasture as they needed on very reasonable terms. For a new loan farm a farmer paid nothing and the quitrent was so trifling, and the government so accommodating and lax with respect to collecting it, that it would not have prevented even the poorest man from becoming a stock farmer.

Moreover, right through the pioneer period great stretches of suitable crown land always lay unoccupied, and if someone was not able to pay the quitrent for a loan farm, he could always lead a nomadic existence on the crown land. Otherwise he could establish himself somewhere without formal title and remain there until someone else took the land by lease. Colonial laws certainly forbade this, but many did it anyhow and no one was ever prosecuted.

In practice the amount of land made available for stock farmers in this manner was absolutely unrestricted. Colonists could settle everywhere that usable pasture was available and apply for farms on loan. Such requests were never denied, otherwise the colonists would have used the land without permission and the Company would have lost the quitrent. The government also never seriously tried to limit the available crown land by preventing further expansion inland or even slowing it down. The colonists simply migrated farther inland when good farms became scarce in occupied areas. And the government, which could not maintain the colonial frontier, had to follow along behind in order not to lose taxation.

The Company's liberal land policy worked against the rise of a white working class at the Cape. As a result of careless land distribution there existed right through the Company period a relationship between the supply of and the demand for crown land that excluded all competition for the cheapest land on the market. This enabled every free laborer to become a landowner and to farm independently.[23]

Had the Cape government made it more difficult for stock farmers to obtain the necessary pasture by setting higher requirements on the initial capital holdings of the applicant and preventing squatting on crown land, many farmers would have been compelled to work as laborers until they possessed enough capital and experience to become successful farmers. If crown land had been more expensive, stock farming would also have made higher demands on the farmers' enterprise, intelligence, energy and business acumen. People who were capable of achieving more than others under the same conditions would have been able to become independent farmers. Those who were unsuccessful at farming would have been unable to begin anew from the bottom up. A certain number of colonists, therefore, would have been obliged to work as wage laborers because they were not able to live independently. Thus, the labor requirements of farmers with adequate capital resources would have been met by white workers, and the higher prices of products as a result of the restrictions on production would have enabled them to pay reasonable wages. Compulsory labor would not have been necessary, and in the interior there would not have been so many poor, small farmers existing, in a way, independently.

Company officials frequently expressed their concerns about the undesirability of migration into the interior, and more than once suggested that agriculture be organized as in the Netherlands so that the poorer colonists could take up service with the more well-to-do. The Council of Policy concluded, however, that no matter how desirable the means were,

> it can be considered as a matter of absolute impossibility, . . . except, no doubt after the passage of a few centuries, when the

uninterrupted increase of inhabitants shall have taken away all
space from the land, so that eventually no corner of fertile
ground is left, this will stir each man to enter employment
rather than going to work on the land and remaining his own
master; when it proves possible to stop the unceasing migration
of the inhabitants into the still unoccupied areas of the interior;
when poverty, and the accompanying shortages in their lives
shall overcome their shame of being servants; when the happy
effects for the inhabitants is achieved through this coincidence
of affairs, to work for which now would be considered a very
futile undertaking by every sensible person who knows this
country and its inhabitants.[24]

The Desire for an Independent Existence

In response to the Burger petition of 1784 the Council of Policy
stated,

Concerning the fear which is shown in the aforementioned address, that the inhabitants in the far-lying lands shall degenerate
into a wild and barbaric folk. We take the liberty of saying very
respectfully that, although what was said in that regard in the
address is not entirely groundless, it is, on the other hand,
nevertheless far from the case that anyone other than the people
themselves are to blame for their degenerating into such a wild
people, because it has not been out of a need to expand stock
farming, but it has been simply and solely the craving of the
inhabitants to be their own masters instead of remaining nearer
to the chief town and serving their equals.[25]

This unsympathetic answer highlights another important factor that worked against the development of a white working class
at the Cape—namely, the desire for an independent existence. This
is not a specifically South African character trait but something
universally human that plays an important role in the economic
development of all young countries. Therefore, it ought to come as
no surprise that the farmers' sons chose to go and farm independently on their own in the interior, as long as it was possible, rather
than to hire themselves out to someone else: to be your own master
is always preferable to working for another. In South Africa, however, special conditions made this attitude more complicated and

kept alive the striving for an independent existence even in cases where it led to a lowering of the living standard.

Poor economic conditions at the Cape made the accumulation of extraordinarily great riches impossible, and the amassing of capital in the hands of certain families was still further counteracted by the fact a farmer generally left his children equal inheritances.[26] In addition, there was little difference in the countryside between the lifestyles of the rich and the poor.[27] A rich man was not differentiated from a poor man on the basis of the clothes he wore, the house in which he lived, and the furniture he possessed, but on the basis of the number of sheep in his fold and the welfare of his children. Luxury and opulence were unknown, even among rich farmers. The cash a farmer took in was usually carefully put away until the need compelled him to use it. The only outward displays of prosperity were perhaps a handsome horse wagon, a span of oxen of a single color, or a number of fine horses.[28]

In these circumstances, during the pioneer period there developed no social and economic class differences between whites in the interior. "They know no other difference, no rank in society, other than that which color imparts to the skin,"[29] explains W. von Meyer. "There is little or no gradation of ranks among the white population," Thompson assures us. "Every man is a burgher by rank and a farmer by occupation."[30] This absence of class distinction, together perhaps with the Christian ideas of brotherhood and equality, was chiefly responsible for the expressly democratic view among the farmers whereby whites were one another's equals.

If one white went to hire himself to another for wages, the social gap that naturally would have developed between the employer and the employee would violate this principle of social equality. More than that, the idea of working for a master offended the sense of Afrikaner independence, for they generally considered themselves totally independent economically as soon as they owned a gun, a wagon, and some livestock. No one in the old days, when game and pasture were still abundant, was so poor that he would not have felt humiliated to be a servant of another.[31] "The farmers are the nobility of this Colony," explained Moodie, "and as long as their children can find a piece of ground to cultivate on their father's estate, they cannot bear the idea of quitting their imaginary rank and losing caste."[32] De Mist also made mention of this peculiar prejudice of the Afrikaner colonists against working for others: "Men send no children from their homes—prejudice prevents the children of the one working in the service of the other. They marry among themselves and thus have to have a 'farm.'"[33]

The Disgrace Attached to Manual Labor

Sometimes one encounters the view that the colonists considered it a disgrace to work and that they were disdainful of manual labor as such. So Baron van Imhoff complained in 1743, for example, that "the majority of agriculturists in this colony are less farmers than they are bosses of plantations and many among them would consider themselves disgraced should they ever have to use their hands for manual labor."[34]

There is not sufficient reason to assume that the older generation of pioneers regarded it a disgrace to perform manual labor. Back then, just as today, it often happened that a farmer and his sons did all manner of farmwork themselves in order to save on hired labor as far as it was possible.[35] No farmer would hesitate to walk behind his own plow or to stand at his own bellows. It is well known that on the Great Trek even girls led the teams of oxen and drove the livestock. The older generation's bias was obviously not directed toward manual labor as such but against the performance of manual labor for a master for fixed wages, whereby you lost your independence and placed yourself in the position of a hireling or servant.[36]

Equality with Slaves and Khoikhoi

This service relationship was made more degrading because in the old days slaves and Khoikhoi, regarded as heathens and members of a lower racial group by the colonists, were mainly the only ones who worked for masters. The general opinion was that if a farmer's son hired himself out to another, then he placed himself by that act on an equal level with slaves and Khoikhoi. The use of slaves and native labor at the Cape thus helped to counter the development of a white working class in so far as it widened still further the social gap between employer and employee, which clashed with our pioneer tradition.[37]

This attitude of the colonists, that we have established from contemporary sources, appears very clearly as well from the well-known petition of 1784. The citizens declared frankly

> that the undersigned do not want to deny that it would certainly provide better employment for various young colonists if agriculture was instituted here in the manner of the fatherland, and men were therefore subservient to one another, but at the same time they believe it justified to comment on how difficult it is to expect this to happen here. That men in the colony around Cape Town work with slaves, and in the interior they mostly work with Khoikhoi. It is quite a different business to work in one's father's house and make a contribution to the income than to perform the same work as a servant boy in the service of another. Our young people would be engaged in a service equal with the Hottentots and slaves, at least so long as the necessary measures are not devised through a clever and careful organization here, and they are gradually led to this.[38]

"Kaffir Work"

There exist no adequate reasons to presume that specific forms of manual labor were considered in the old days as socially inferior because they were generally done by nonwhites. In modern times there is certainly a tendency among the poorer classes of whites to view certain sorts of work as "kaffir work"—that is to say, work that is not suitable for a white. Especially where gang work is used, the unskilled labor is regarded not only by the poor whites themselves as "kaffir work," but the entire white community views it as such. Sometimes the bias against such work rests principally on the fact that it is predominantly performed by blacks. Sometimes it rests on the fact that wages paid for the work are considered too low for a white. This prejudice is presumably something from a later date, however, as the name clearly indicates.[39]

Whites in Trades

It was not only the labor requirements for agriculture that offered colonists a potential livelihood in the old days; there was constantly a great demand for tradesmen as well. Under pioneering conditions up-country the need admittedly was not very great: partly as a result of the poor levels of prosperity and low living standards of the farmers; partly as a result of the diffusion of the rural population, which made the division of labor difficult. In the older areas of settlement there was, however, a greater need for tradesmen because the people were more prosperous and lived more closely together.

Unfortunately, trades at the Cape fell into the hands of slaves relatively early on. The majority of slaves were thoroughly trained for special sorts of work and their masters then hired them by the day, by the week, or by the month as bricklayers, cabinetmakers, shoemakers, tailors, and so forth. In and around Cape Town there were a number of citizens that lived off this type of work by their slaves. "However harmless it may seem in the abstract," declared de Mist, "it is nevertheless still a great evil, that a white is ashamed to appear himself in public with a leather apron, saw, and trowel and participate in this work."[40] In the Burgher Petition of 1784 as well there was the complaint that "the faculty of earning money through the work of slaves prevents many a European hand from practicing a trade."[41]

In this connection the colonists' prejudice was obviously not aimed against the sort of work that a tradesman had to perform. During the period of the self-reliant farm life, nearly every pioneer was obliged to be a bricklayer, carpenter, smith, and shoemaker. No farmer would have been ashamed to be found at home working with a hammer, a plane, or a trowel. Frequently in areas of older settlement farmers' sons even specially acquired one or another trade. This was normally used, however, only as a means to obtain capital to buy a farm and to begin independent farming.[42] It hardly ever happened that a white pursued a trade as his life's work "out of fear of being thereby disgraced and placed on an equal level with a slave."[43]

Even imported craftsmen never practiced their trade for long. In 1785 the Council of Policy complained,

> that for a long time experience has shown how such people, once they are discharged here from the Company's service, and given freedom as a citizen, or have arrived as a free burgher from the Fatherland, while pretending that they shall practice

the one or another occupation for the convenience of the community, very seldom remain continually bound to the same trade. A short time after being discharged from the Company's service, they change over to other trades or jobs, requiring much less physical labor and consequently affording an easier existence. In following their subsequently accepted calling, they demand just the same new craftspeople for their benefit, as for which they themselves acquired their freedom as burghers, or for which they have come over here from the Fatherland.

That as the cause never ends by which the extension of this colony has come to take place, a complete conviction of the disadvantage thereof has led to the government here to be ever more reluctant in the emancipation of Company servants as free burghers, to which they consequently do not proceed except in the utmost necessity.[44]

These circumstances led de Mist to express the opinion that it would be good to set up guilds at the Cape "and secure civic honor and pride, not in order to place the industry there in fetters, but to break the white inhabitants of the notion that manual labor is shameful and because regulations regarding the present unconscionable and arbitrary daily wages could then be enacted."[45]

The Tradition of Independent Farming

The fact that *Boere* [farmers] became the national term for our people was by no means accidental. For centuries every Afrikaner was indeed a farmer and farm people were the aristocracy of our country.

Without doubt this was primarily the result of three factors that are linked closely together: the liberal land policy of the Company at the Cape, the universally human desire for an independent existence, and the fact that at the Cape much use was made of slave and native labor.

A closely related factor was the group isolation in which farmers' sons grew up. They only came into contact with farmers and received no vocational guidance. Thus, they bore little knowledge of other callings or spheres of work outside of farming and had no opportunity to learn their value. In addition, they lacked the opportunity to obtain the necessary training for other professions.

Each farmer's son was compelled in very great measure, therefore, irrespective of his aptitude, to become a farmer.[46]

Besides, it was quite natural that children, who had very great respect for their parents, would have been inclined to follow in their fathers' footsteps where it concerned the choice of a lifelong career. This was directly promoted moreover by the tradition among farmers of marking lambs for their children, as well as by the fact that children normally received their share of the inheritance in livestock. Thus, there developed over time a deeply rooted tradition that sooner or later a farmer's son must also become an independent farmer.[47]

This attitude appears very clearly from the report of a interview that Moodie had in 1834 with the migrant farmers on the northeasterly border of the colony:

> Blamed them for making all their children graziers, and ascribed to this custom many of the inconveniences to which they were subject. P. Botha (later to Natal) said "aye! that's the point on which I would like to hear what you can urge; we are stupid enough, but we see that; but, tell me, who will employ our children, if we teach them trades? Not the Boers, for they are Jacks of all trades. Among the English it is different, because they learnt different employments before they came hither, and are accustomed to buy what they want, which we will not do, while we can make any shift, or do without; we *must* keep sheep, and seek pasture, as far as government will allow—or starve."
>
> On asking—"But why not rather serve one another, than accumulate stock without land?" another replied "Because every child is a Boer, and gets his share of the paternal or maternal inheritance in stock, and in what country will people serve for hire, if they can live their own masters?"[48]

This tradition meant that stock farming in the rural areas for many years expanded in the same proportion as the population increased. It caused a constantly growing demand for new land, which led to an uninterrupted migration inland.

NOTES

1. C 1283, Petitions, Reports, etc., Burgher Petition of 17 February 1784.

2. AR, Accession 157 (Nederburg), van Ryneveld, Inquiry, 30 April 1805, p. 43.

3. PSB, Mss. Germ. Fol. 879, de Mist, Opinion (1805), pp. 95–99.

4. See chap. VI.

5. W. J. Mackrill, Evidence, 15 January 1827, in G. M. Theal, *Records of the Cape Colony, 1793–1831* (Cape Town, 1897–1905), XXX:407.

6. C 171, Resolutions, 19 July 1786, pp. 459–61.

7. J. Barrow, *Travels into the Interior of Southern Africa* (London, 1806), II:28; R. J. Cleveland, *In the Forecastle, or Twenty-five Years a Sailor* (New York, [ca. 1842]), pp. 25, 63; W. J. Burchell, *Hints on Emigration to the Cape of Good Hope* (London, 1820), p. 100–01; M. D. Teenstra, *De Vruchten mijner Werkzaamheden gedurende mijne Reize over de Kaap de Goede Hoop, naar Java en terug, over St. Helena, naar de Nederlanden* (Groningen, 1830) I:375; C. Latrobe, *Journal of a Visit to South Africa in 1815 and 1816 with some account of the Missionary Settlements of the United Brethren near the Cape of Good Hope* (London, 1812), p. 93; Commissioners of Inquiry: Report II, Upon the Finances at the Cape of Good Hope, 6 September 1826, in Theal, *Records*, XXVII:424.

8. G. Thompson, *Travels and Adventures in Southern Africa* (London, 1827), II:133.

9. Swellengrebel Family Archives, 10 January 1780. [Trans. note: This quotation is from a letter from an unidentified author to Hendrik Swellengrebel, Jr., and may also be found in the collection of letters published by the Van Riebeeck Society, *Briefwisseling van Hendrik Swellengrebel Jr. oor Kaapse Sake, 1778–1792*, VRS, series II, vol. 13, p. 100 (in the original with a shortened English summary on p. 326).]

10. Barrow, *Travels*, I:80; Janssens to de Mist, 16 April 1803, in G. M. Theal, *Belangrijke Historische Dokumenten over Zuid-Afrika* (Cape Town, 1896, 1911), III:211–12.

11. Cf. H. Lichtenstein, *Reisen im südlichen Africa in den Jahren 1803, 1804, 1805 and 1806* (Berlin, 1811–1812), II:57.

12. A. W. Dryson, *Tales at the Outspan; or, Adventures in the Wild Regions of Southern Africa* (London, 1865), p. 24.

13. AR, Accession 169 (Nederburg), van Ryneveld, Brief Observations 19 March 1804; Swellengrebel Family Archives, Cloete, Journal, p. 27.

14. Cf. E. G. Malherbe, *Onderwys en die Armblanke* (Stellenbosch, 1932), pp. 23–24.

15. Commission for Stock Raising and Agriculture, dated 20 November 1805, in Theal, *Belangrijke Historische Dokumenten*, III:426. Cf. also Swellengrebel Family Archives, Swellengrebel to De Gyselaar, 26 June 1783, and R. W. Wilcocks, *Die Armblanke* (Stellenbosch, 1932), p. 63.

16. J. F. W. Grosskopf, *Plattelandsverarming en Plaasverlating* (Stellenbosch, 1932), p. 110.

17. Cf. Wilcocks, *Armblanke*, p. x.

18. Commission for Stock Raising and Agriculture, dated 20 November 1805, in Theal, *Belangrijke Historische Dokumenten*, III:426.

19. O. F. Mentzel, *Vollständige und Zuverläszige . . . Beschreibung . . . des Guten Hoffnung* (Glogau, 1785, 1787), II:169–70.

20. G. Foster, *A Voyage round the World* (London, 1777), p. 76; Barrow, *Travels*, II:410.

21. L. Fouche, *Het Dagboek van Adam Tas* (London, 1914), p. 22.

22. J. W. D. Moodie, *Ten Years in Southern Africa* (London, 1835), II:5; T. Pringle, *African Sketches* (London, 1834), p. 182; J. Holman, *A Voyage round the World* (London, 1834–1835), p. 308; Mentzel, *Beschreibung*, II:171.

23. Cf. E. G. Wakefield, *A View on the Art of Colonization* (London, 1849), pp. 325–39.

24. C 171, Resolutions, 19 July 1786, pp. 459–61.

25. C 1813 and 1814, Letters Disp., C. J. de Graaf to Kamer Amsterdam, 19 April 1786.

26. Moodie, *Ten Years*, II:5.

27. Theal, *Belangrijke Historische Dokumenten*, III:426.

28. Moodie, *Ten Years*, II:5; A. Sparrman, *Reize naar de Kaap de Goede Hoop, de Landen van den Zuid Pool, en rondom de Waereld* (Leiden, 1786), II:594.

29. W. von Meyer, *Reisen in Süd-Afrika wärend der Jahre 1840 und 1841* (Hamburg, 1843), p. 82. Cf. AR, Accession 157 (Nederburg), van Ryneveld, Inquiry, 30 April 1805.

30. Thompson, *Travels*, II:275.

31. Ibid.; J. MacGilchrist *The Cape of Good Hope (By a Traveller)* (Glasgow, Edinburgh, London, 1844; 2d imp., London: Frank Cass, 1965). Cf. also Grosskopf, *Plattelandsverarming*, pp. 166–67; Wilcocks, *Armblanke*, p. 55.

32. Moodie, *Ten Years*, II:27.

33. PSB, Mss. Germ. Oct. 275, de Mist, Memoranda van der Kaapsche Landreize (1803).

34. C 2357, Instructions from van Imhoff (1743), p. 29.

35. AR, Accession 157 (Nederburg), van Ryneveld, Inquiry, 30 April 1805, p. 41; Grosskopf, *Plattelandsverarming*, p. 166.

36. Cf. Grosskopf, *Plattelandsverarming*, p. 170.

37. PSB, Mss. Germ. Fol. 879, de Mist, Observations (1805), p. 118; and PSB, Mss. Germ. Fol. 885, Oud-Fiskaal Boers, Reflections (1803); AR, Kol. Arch. 4365, de Mist to Janssens, 26 November 1803, p. 40; Swellengrebel Family Archives, Swellengrebel to De Gyselaar, 26 June 1783; C 2357, Instructions from van Imhoff (1743), p. 29; C 1813 and 1814, Letters Disp., Cape Government to Patria, 19 April 1786, p. 521; cf. Grosskopf, *Plattelandsverarming*, p. 166.

38. C 1283, Petitions and Reports, Burgher Petition of 17 February 1784, p. 59.

39. Grosskopf, *Plattelandsverarming*, pp. 36–43; Wilcocks, *Armblanke*, pp. 53–58.

40. PSB, Mss. Germ. Fol. 879, de Mist, Observations (1805), pp. 106, 109.

41. C 1283, Petitions, Report, etc., Burgher Petition of 17 February 1784.

42. Moodie, *Ten Years*, II:27; Thompson, *Travels*, II:275.

43. PSB, Mss. Germ. Fol. 879, de Mist, Observations (1805), p. 111.

44. C 169, Resolutions, 11 November 1785, pp. 709–10.

45. PSB, Mss. Germ. Fol. 879, de Mist, Observations (1805), p. 111.

46. Cf. Malherbe, *Onderwys en die Armblanke*, pp. 132–33; J. R. Albertyn, *Die Armblanke en die Maatskappy* (Stellenbosch, 1932), p. 23.

47. CO 41, Court of Justice, General Report of the Commission of Circuit (1811–1812); Thompson, *Travels*, II:132; MacGilchrist, *Cape of Good Hope*, p. 30.

48. Quotation taken from D. Moodie, "Saxon Nomads," in *South African Annals* (Pietermaritzburg, 1855), p. 36.

V

Living Conditions in the Country Areas

Economic Conditions

Throughout the eighteenth century and the early years of the nineteenth century the farmers in the interior of the Cape Colony led a fairly poverty-stricken existence in nearly absolute isolation. Distances were great and all sorts of natural obstacles made communication with the capital and the conveyance of products to the market difficult. Cape Town was surrounded by extensive flats of deep sand and was furthermore shut off from the interior by high mountains, rising steeply on the coastal side. It was often impossible to cross over some of the natural mountain passes with a loaded wagon. Going up the mountain the driver had to rest his oxen every few minutes. Going down there was always the danger the wagon would run over the rear oxen. Even brake chains and brake shoes did not always guarantee a safe downhill journey.[1] One could barely travel down Mostertshoek, the Witsenberg and the Skurweberg, even with an empty wagon, without having an accident.[2] As a result the farmers from the Cold Bokkeveld who brought their peas and beans to the Cape had to unload their wagons on the mountains, load the produce on packhorses or portage oxen, drive down the passes with empty wagons, and reload them again at the foot of the mountains.[3]

Over such routes a wagon could make only a few journeys to the Cape, and transport was exceedingly expensive considering that in the old days everything had to be transported by ox wagon for want of navigable rivers. This is clear from the following report of Governor Janssens:

The transports are worth seeing. One wagon hauls not above, but well below, 2,000 lbs. Ten to twelve oxen pull this and not infrequently they are followed by an equal number of oxen for leaders or more often for exchange. Among the various wagon transports we came across there were three together, made up of as many as 18 to 20 persons of various colors and years, even children at the breast, and more than 70 oxen; and taking nearly 200 hours to convey in this manner a load of 5,000 lbs., that was of little value except for the livestock.[4]

Save for the work and the wear and tear on the wagon and harness, there were not many expenses connected with a journey to the Cape in the old days. It was not necessary, for example, to purchase fodder for the oxen as the government had reserved out-span places everywhere along the main road especially for use by travelers. In addition almost no farmer would deny a traveler a place to outspan if he would just ask for it and not use it as his right. On the road a farmer was just as much at home as in his own house. He took along dried and cured meat and bread in the wagon chest. If he was not driving slaughter cattle, he could pur-chase an animal to slaughter anywhere along the road. Now and then he outspanned in order to allow the oxen to graze a little. Then a fire was lit, water boiled for coffee, and while the men sat and smoked their pipes, the women prepared a meal.[5]

All sorts of frustrations often awaited the travelers, however. Outspan places were frequently overgrazed, and during droughts it was hard to find enough fodder along the road for the animals. Washed out fords and full rivers sometimes held up wagons for days and weeks at a time. A journey to the Cape was also accom-panied by a number of other problems for the frontier farmer. It was generally not very safe to leave the wife and children alone back on the farm. If everyone was away from the house, then the work might come to a standstill. The herders could let the livestock loose or neglect the lambs; the Bushmen could steal the livestock or burn down the farm. Finally, the journeys to the Cape were time-consuming. Even if all went well, the ox wagon still moved very slowly. Without overworking the livestock a person could travel with a loaded wagon in one day a distance that took four hours on horseback, taking about eight hours in order to complete such a stage of the journey. By alternating the extra oxen in the team, however, a person could, if the need was pressing, do two stages of the journey in twenty-four hours, thus covering about forty-eight miles.[6]

Under these conditions it is understandable that a journey to the Cape could not be undertaken very often. Even farmers who did not live very far from the Cape usually visited the capital only once a year. These visits became even less frequent as they migrated deeper into the hinterland. The trip from Camdeboo to the Cape and back lasted from two to three months. Farmers living at an even greater distance often had to be away from their farms between four and five months if they wanted to make a journey to the capital. Consequently the farmers in those remote parts did not visit the Cape more than once every two, three, or even four years.[7] Many visited the Cape only once in their lifetime: when they went to the capital to get married! Usually the farmers seldom stayed at the Cape longer than a day. They outspanned by the Salt River the evening before and rode into town at daybreak. They then settled their affairs as quickly as possible so they could get away from this place where townspeople regarded them as country bumpkins and where there was no grazing for the oxen.[8]

In the interior itself no country markets or commercial centers of any significance developed during the eighteenth century and the first years of the nineteenth century. Apart from a very limited barter in special products like brandy, tobacco, dried fruit, and grain, which did not even occur on a regular basis, there was almost no trade among the different districts. To a certain extent transportation problems and long distances made inland trade very difficult, and because the grain and garden farmers generally kept livestock, there was even among them little need for the stock farmer's products.[9] Thus Cape Town remained the only market for the inland farmers' goods.

Because this distant market was virtually inaccessible in the old days, when farmers still had to transport their own goods, slaughter animals were the principal product delivered to the market from the interior. Slaughter animals transported themselves to the Cape, and in fact it was not even necessary for a farmer to drive his livestock to market. Livestock was sold to itinerant butchers' servants right on the farms. Payment was not made in hard cash but with signed bills of exchange on which the butchers' servants simply had to fill in the specific amount. The money was then paid out when the bills were tendered later in the Cape. Of course a farmer could not always wait for his money until he again made a journey to the Cape, but that was not an insurmountable difficulty. "Butcher's notes" were commonly certified by the fiscal as genuine in order to avoid fraud, and often circulated in the countryside as currency before they were presented at the Cape for payment.[10]

This system was very convenient and saved the farmer many problems, but it was not always the most suitable arrangement for the establishment of a regular market. The farmer was dependent on the arrival of the butcher's servant. If the buyer did not come when the farmer's livestock was fat, then he was stuck with it. And then if the buyer finally turned up one day the livestock were perhaps no longer prime for the market. In addition, the butchers' servants often made life easier for themselves by purchasing all the livestock they needed from a few rich farmers. As a result, the more remote farmers, particularly when slaughter animals were abundant, could not always count on a regular market for their livestock.[11]

Grain cultivation at the Cape during the eighteenth century was not amenable to extensive development because of limited market opportunities.[12] Besides, most of the interior was not particularly well suited for it. Even where grain could be cultivated, there existed little inducement to sow more than what was necessary for one's own use. The time and expense associated with the slow transport by ox wagon quickly made competition with the farmers living close to the Cape impossible.[13] Accordingly, the grain farmer remained confined within a relatively small area of the southwestern Cape. Cloppenburg reported in 1768 that the last farm from where it was still economically feasible to transport grain to the Cape was situated where Tijgerberg ended and the Hessequas Mountains began.[14]

The efforts made toward the end of the eighteenth century to produce grain for the Cape market in the Mossel Bay area were hopeless failures.[15]

Other agricultural products that had greater value than grain in relation to weight, like peas and beans, could be conveyed to market over longer distances, but there were naturally limitations even here.

What is more, the interior districts did not yield many other products that could be brought to market on a regular basis. Soap could be cooked when the livestock were fat and where the lye bush was available for lye. The farmers in predominantly cattle-farming districts, in the eastern part of the colony, also produced salted butter for the market. Once or twice a year the farmers living near the Cape brought in a load of butter. It was usually sold for about three to six stuiwers per pound. Farmers who owned only average quantities of livestock and land (Sparrman assures us) could bring from 1,200 to 3,000 pounds of butter to the market yearly.[16] Distance soon made production for the Cape Town market difficult, however. Moreover, making butter played a less important role in districts where sheep and goats were predominant. A small amount of aloe came from Mossel Bay, and farmers who had success on the hunt could sell ivory, horns, and animal skins in Cape Town.[17] The plan to establish a forestry settlement in Outeniqualand never materialized.[18]

Under these circumstances, the conveyance of products to Cape Town never assumed great economic importance. Often a load of products from the remote interior districts consisted of little more than what was necessary to cover the expenses connected with the journey.[19] In fact, the farmers often did not have as their primary reason for making the long journey to Cape Town the marketing of their products. There were other reasons that often made it necessary for frontier farmers in the eighteenth century to visit the capital. They had to arrange the government lease for their farms and pay the quitrent. They had to go there to contract a marriage and to have serious disputes settled. They also had to exchange butcher's notes and purchase items their farms did not produce, like ammunition, agricultural implements and other iron-ware, fabrics, groceries, tobacco, and perhaps a keg of brandy.[20]

With the passage of time the necessity of traveling to the Cape in order to take care of shopping needs disappeared. Along the southeast coast and in the hinterland, small villages developed where merchants settled. The itinerant trader also gradually made his appearance in the hinterland. By the time of the Great Trek the itinerant trader already played a very important role in the economic life of the frontier districts, regularly traveling around among the farmers with two or three wagons or carts loaded with clothing and groceries. Itinerant traders could usually supply nearly everything the farmers needed, and also took orders for their next visit.

The itinerant trader normally let the farmers know beforehand that he was coming and also what route he was planning to follow. And while the news was spread in every direction that he was approaching, he slowly moved from one place to another, frequently staying over a few days at a centrally located site. This gave the farmers in the neighborhood the opportunity to come and do their shopping. At such a stand buck sails were stretched between the wagons, chairs were put out for the buyers, and the itinerant trader unpacked his goods. Thus did, among others, Willem Robinson, the father of the millionaire with the white helmet, trade in the Koonap before the Great Trek. The arrival of the itinerant trader, who was generally an old acquaintance, was always a great event in those days, because he brought not only necessaries, but often mail and always at least some news.[21]

Apart from the butcher's servant and the itinerant trader, the frontier farmers existed nearly economically independent of the outside world. Isolated in the interior each patriarchal farm family formed a more or less closed and self-sufficient economic unit. Production for a faraway market that was nearly unreachable and very limited soon became less important than production for one's own use. And while there was little surplus produce to bring to market, it was difficult to satisfy any needs that the family could not take care of themselves. Consequently, frontier farmers had to make do without store-bought goods as much as possible and in reality, with a few exceptions, limited their needs to what could be made and produced on their farms.

In addition, the division of labor, whereby different types of work were entrusted to separate people, each one with a special ability and training, was nearly unknown; among the thinly scattered pioneer population each individual family formed a self-sufficient production unit. The pioneer in the interior had to build his own house and more often than not make his own furniture. He chopped his own wood, at times constructed his wagon with his own hands, and drove his own transport. He even worked the skins of the wild game he shot and the hides of the livestock he slaughtered, and for tanning he used the bark his children and slaves had gathered. He braided his own leather straps, with which he replaced all kinds of rope work. He made shoes for himself and his family and used thongs that he braided himself. In short, a farmer in the old days was not only a farmer but also mason, carpenter, smith, wagonmaker, shoemaker, and transport driver. The wife acted as tailor and seamstress for the family, cooked soap with lye from the lye bush, and made candles from the suet of animals—tallow-dipped or cast candles, with wicks of twisted linen. Thus, isolation in the interior forced the pioneers to be versatile and technically resourceful, which facilitated self-sufficiency and independence from the outside world.[22]

It is very difficult to form a clear understanding of the prosperity and living standard of frontier farmers toward the end of the eighteenth century and the beginning of the nineteenth century. Nevertheless, from the sources of that time it seems very obvious that the general level of prosperity of the farmers then was much lower than it is today. Poor marketing opportunities and bad roads restricted the economic development of the interior and kept the pioneers on the frontier poor. The unsettledness of pioneer life also made the accumulation of riches difficult. For many years the Xhosa on the eastern border again and again destroyed a considerable portion of the pioneer population's collective property: it was not unusual for a farmer to have his farm plundered four or five times in a lifetime. On the northern border Bushman thievery ruined many farmers. Numerous families frequently found themselves suddenly reduced to poverty.[23]

Throughout the pioneer period, when everyone could still receive crown land for free, it was certainly much easier than today to enjoy an independent existence as a stock farmer. But to make money from it in the old days was very difficult. The general scarcity of money in the rural areas during the eighteenth century is well illustrated by the problems that farmers experienced in trying to raise the low quitrent money in order to pay for their loan farms. Petitions for the stoppage of quitrent and remission from outstanding debts continually streamed in. Throughout the eighteenth century the Company government never ceased complaining about the large sums of quitrent for which the farmers were in arrears. And in most cases this resulted not "from willfulness but truly from poverty and lack of money," as the Council of Policy concluded in

1740 from the Stellenbosch landdrost's answers to questions about this matter. [24]

The pioneers in the interior thus had to live a simple life out of sheer necessity and make do without many of the conveniences of civilized life. Luxury was altogether unknown in the rural areas. [25] Even farmers who enjoyed relative prosperity according to the ideas of their time and could live generously, often made do with very little. While the inhabitants of Cape Town and its environs, mutually envious and very aware of the preservation of class distinctions, were inclined to live above their means and tried with elegant homes and fine furniture, bountiful tables and smart clothing to wipe out the distance between themselves and their fellow citizens or high-ranking officials in the hinterland, [26] there was not a great deal of difference in lifestyle between rich and poor. "It is not their clothing, not their furniture, but rather the number and healthy nature of their cattle, and above all the strength of their draft oxen, over which the farmers compete with each other for prestige," [27] Sparrman remarked in his well-known travelogue. Mentzel also mentioned the rural farmers, "who should rather more deservedly be called cattle herders than farmers." He then explained, "Among these there are equally rich and poor; but the former, with all their overabundance of cattle, must live a very wretched existence just as the poor, and are very badly in want of everything that serves as a comfort to human life." [28]

The daily account of Hendrik Swellengrebel's tour in 1776 throws an interesting light on the economic condition of eighteenth century frontier farmers. Throughout their journey the party of travelers stopped and attempted to gather information about the farmers' means of livelihood and economic prospects. Nowhere, however, did the information appear to be very rosy. Conditions in Camdeboo were described as follows:

> Anyone who keeps these farms (which because of the scantiness of the grazing, cannot possibly be nearer together) for purposes other then the breeding of livestock will not make much profit. Different people have settled themselves here and could have a very good life, if transport to Cape Town was not so difficult and costly. *The majority here now live not much better than the Hottentots.* [29]

On the eastern frontier economic conditions were not much better:

> along the Great and Little Fish rivers up to here lay fourteen to fifteen farms, yet on the majority the occupants live a very indifferent existence. This is equally evident from the following statement of one F. Botha, living around here, namely:

> He sells annually at the most 200 hamels at 5 schellingen ƒ150

> He has produced no butter now in 2 years, yet if he were to acquire cattle again, he could make 1,000 lb. at 3 st ƒ150

> He can still not make soap, because here abouts no Kanna bushes grow; but he hopes to be able to make preparations, although he will then have to haul in wood and twigs from afar.

> Against this he harvests no seed grain and therefore has to buy his bread, that calculates to 50 muids a year at a Ducaton ƒ180

> Then he has to have clothes and the quitrent of 24 Rixsdollars. He has no hope of being able to pay for oxen. His riches consist of 500 sheep and a pair of cattle with one wagon.

> So it is with others who went in search of more room. I have not seen that they are better off, and the majority have even barely enough to eat. This is also the reason why on various farms several families live together in partnership, as at old Koekemoer, where three married couples with their children live in the house described before, and at K. van der Merwe's and others more. In general, the colonists living in and nearby the principal towns are not very industrious; those living at farther distances away are even less so; they adopt fully the character of cattle herders and wild game hunters.[30]

Complaints about poor economic conditions and the poverty of the rural population were universal by the end of the eighteenth century. And during the first quarter of the following century the

situation evidently did not improve. Just before the council's abolition in 1827, the Graaff-Reinet landdrost and heemraden wrote to
the colonial secretary,

> We have before breaking up one more sacred duty to perform
> to the public, in the execution of which the liberal spirit of His
> Majesty's Government, and His Honor's known love of truth
> and Justice dissipate every fear of giving offense. This duty
> consists in a representation of the State of poverty and distress
> to which the inhabitants of this district are reduced, and their
> utter inability to meet the taxes proposed by His Majesty's
> Commissioners. Every species of trade and traffic is completely
> at a stand; the flocks in the possession of the farmers, far from
> being a source of revenue, are a burden to the proprietors, as
> they have no market whatever for their produce, the most tri-
> fling loan is not to be procured on any terms, and money seems
> to have totally disappeared, so that we look with terror to the
> period when the several contributions shall be called for.[31]

Housing in the Country Areas

It is understandable that in the early days housing in the interior
would be much poorer than in Cape Town and the areas around
the capital. The ostentatious Capetonians gladly showed off their
beautiful, large houses. Often the houses were so enormous the
entire building could not be occupied and some of the rooms were
left unfurnished. "Home construction here is not only a favorite
pursuit, it is a passion, a madness, a contagious frenzy, which has
infected most everyone,"[32] explained De Jong, who visited the Cape
in the second half of the eighteenth century.

Roomy and convenient dwellings were also to be found in the
agricultural regions near the Cape. Some of these beautiful old
homes are still to be seen today. They were usually built according
to the French-Dutch style of the eighteenth century, with graceful
gables and green painted window frames against whitewashed
walls. In the windows there were elegant sashes with a great number of small panes. Closing over the windows were green louvers
or, if not, then wooden blinds that folded inwardly. The thick walls,
heavy top beams, sturdy loft boards, solid doors, and great hinges

created the impression of overall solidity and durability.[33] In addition to the main house on a Boland farm there were also dwellings for slaves and laborers, storehouses and barns, wagon houses and workshops. On large farms one could often find eight, ten, or even more buildings of various sizes. At times the outbuildings were in the form of a regular square arranged in front of, or behind, the main house. The yard was normally enclosed by a white ringwall. J. A. de Mist wrote that such a farm with its outbuildings and ringwall reminded one of a medieval cloister.[34] Some of the Boland farms even had the appearance of small villages, for on such a farm there lived, besides the owner and his family, a number of slaves and Khoikhoi and their families. At times there must have been more than a hundred mouths that had to be fed daily.[35]

In predominantly stockfarming districts the need for outbuildings was naturally much smaller. Also, the great white stucco farm houses amidst green oak trees remained confined to the regions surrounding the Cape. The farmhouses became smaller, simpler, and more shabby as an eighteenth-century traveler ventured farther away from the capital. In 1776 when the colony stretched out as far as the Sneeuberg and the Fish River, Hendrik Swellengrebel visited the eastern and northeastern frontier districts. The first white stucco house he saw after crossing the Hex River was on his return trip near Mossel Bay. The last "tolerably good house" that he encountered on his journey to the border was that of Wouter de Vos at the Hex River. The first he saw after that belonged to Jacob Kok on the Great Seekoei River, near Outeniquabos, where he stopped on his return journey. In an easterly direction one could still come across tolerable houses nearly as far as Mossel Bay. But what one saw farther to the east "were nothing but miserable hovels."[36] Van Plettenberg, who also visited the frontier districts, supplied similar information two years later. To the east of the Hex River and the Gamtoos, he explained, he found not a single decent house.[37]

By the end of the eighteenth and the beginning of the nineteenth century, the typical country farmhouse, even in areas of older settlement, was very small and uncomfortable according to modern standards. Generally it was around thirty to forty feet long and about twenty feet wide.[38] The first room that the traveler entered was a relatively roomy and airy front room. On both sides of that were two smaller rooms serving as bedrooms and separated from the front room by inner walls that reached up to the beginning of the high-pitched roof. Very often the front room served as sitting room, dining room and kitchen, but the better class of farmhouse usually had a separate kitchen. As a general rule it was built behind the house and connected by a door to the front room. At times cooking was done in a separate kitchen built near the main house.[39]

In the frontier districts the farmers had to be satisfied with even worse housing. Here houses often consisted of only two rooms. One served as sitting room, dining room, and kitchen. The other was used as a bedroom for the entire family and as a storage room. There was no spare room. Unexpected guests and travelers that had lost their way had to be content with a corner of the front room. And often there were a number of Khoikhoi sleeping in the other corner of the room around the fire, while cats, dogs, and even chickens supplemented the company still further.[40]

In regions of more recent settlement single-room cottages were commonly found. In 1778 O. G. de Wet stated that the houses in the Sneeuberg "nearly all comprised a single, low-walled room, without any privacy, loosely covered with a sort of coarse reed, and serving at the same time as a granary and storage for other goods."[41] Living conditions in Camdeboo, only just then colonized, were not much better. Descriptions appearing in the daily account of Swellengrebel's journey certainly do not paint a very attractive picture. The average house in Camdeboo consisted

of a wall of clay, raised 3 to 4 feet high, with a reed roof above, no separation of rooms, without a chimney, the smoke pulled through a hole in the roof, a door of common-reed fastened with a rope, a square hole for a window. The bedsteads are separated from each other by a little Hottentot's mat; consequently the sleeping arrangements are very cozy! The floor is clay mixed with dung and upon this everything is piled together: butter, milk churn, freshly slaughtered cattle, bread, etcetera, while chickens, ducks, and young pigs all swirl around as in a menagerie, and pigeons actually nesting on the roof. . . . In these sheds, that were hardly 40 feet long and 15 wide, two to three families on some farms keep house with their children. Cleanliness was therefore not great. . . . I found but two houses that were decently constructed and tidy, yet were fairly far from luxurious.[42]

Captain Paravicini di Capelli, who visited the Sneeuberg in 1803 together with Governor Janssens, gave the following description of a typical house in the Sneeuberg in his interesting travelogue:

A farm consists of various buildings, some as storage for this and that and some as dwellings for servants and others. I shall describe here only the best, the master's residence, as it is in the summer. The walls all around are in a rectangular shape, do not even reach five feet high, and consist of the same crumbly mud as the floor. On these walls rests a pathetic reed roof, on rough-hewn tree branches of various thicknesses and bound together with skin thongs. The door is a hole of usual width and has the height of the mud wall. The two windows are simply smaller holes, over which are placed planks bound with branches to protect against the wind. There is no chimney, because the so-called cookhouse is a separate hut. According to custom, we built a fire on the floor in the middle of the room, and the smoke found an exit through the poorly sealed holes that were windows, and through the openings in the pathetic roof; not, however, without dreadfully annoying the company. The leaking roof and the cold, inclement night prevented us from sleeping, and lengthened the season's already too long night.[43]

About the same time Barrow described the pioneer dwellings at Algoa Bay, which were similar to those in Paravicini's report,

The miserable hovels in which the graziers live are the pictures of want and wretchedness. Four low mud walls, with a couple of square holes to admit the light, and a door of wicker-work, a few crooked poles to support a thatch of rushes, slovenly spread over them, serves for the dwelling of many a peasant whose stock consists of several thousand sheep and as many as a hundred head of cattle.[44]

In parts of the colony, above all in the Roggeveld, the walls of the houses were usually built with slate and clay mixed together.

This formed a solid, cool wall.[45] In other areas raw bricks or sod were often used for construction.[46] Plank floors or floors of square Javanese stone—which was often found near Cape Town—were not to be found in the interior. Earthen floors were the general rule here until the time of the Great Trek. Anthills were preferred for this, but where they were not to be found, ordinary clay was brought in for it which also made a good floor when properly tamped down. In order to keep such an earthen floor hard and to prevent it from becoming kicked up or dusty, it was smeared daily with cattle dung, thinned with water. The floors were often covered further with skin mats, which under the circumstances were rather good carpets. On occasion the traveler accustomed to better living conditions found the dung-smeared floor a little strange at first, but very quickly he testified to its effectiveness. Even Commissioner General de Mist, only just arrived from stately Holland, later sang the praises of the dung-smeared floor. He wrote in his interesting travel journal, "During our land journey we became so used to these floors, that the discovery of their utility in this climate makes us prefer to have them rather than the finest Asiatic or European carpets."[47]

The galvanized iron roofs of today were still unknown in the interior at the beginning of the nineteenth century. Then houses were often covered with straw, but many times with reeds, bulrushes or other shrubs as well. These rested on a framework of wood or bamboo and were bound tightly together with leather thongs. There was normally never a loft so the roof was visible from the inside. The top beams were used as a hanging place for freshly slaughtered sheep, cured and dried strips of meat, corncobs, tobacco, and all sorts of household articles. On the inner walls, which did not always reach to the roof, the housewife stored her soap.[48]

Transporting glass over untrodden paths and natural mountain passes with an ox wagon was nearly impossible. Accordingly, glass windows became more scarce as one traveled a few days' ride from Cape Town. The farther one traveled inland, the less likely one was to find glass, save perhaps among only the very wealthy farmers.[49] The square holes that were left in the walls in order to let in light and air were boarded over by whatever was near at hand in the absence of the necessary planks. Sometimes a frame was used with white linen stretched over it. At times a piece of hard, un-tanned hide did the job.[50] This state of affairs could still be found in the northern frontier districts until the time of the Great Trek. Farm houses nearer to the coast, however, and even houses on the eastern frontier, all had glass windows by 1838.[51]

There were a number of reasons for the poor quality of housing existing in the countryside. Farmers in the most isolated regions of the interior were often compelled to be satisfied with poor housing because under the circumstances nothing better was possible. Craftsmen were scarce and, besides, these farmers did not always have the money necessary to hire workmen. Building materials were expensive and difficult to obtain. Consequently, nearly every pioneer was obliged to build his own house without the help of craftsmen with whatever materials he could find in his immediate vicinity.

The lack of good wood and iron throughout most of the interior made the construction of large houses very difficult. Only with much effort could the Onder-Bokkeveld's farmers readily obtain good wood in the Cedarberg and alongside the Hantams River. Consequently the houses in this region were generally larger, more comfortable, and better built than the houses in the other field-cornetcies of the north-west.[52] In other field-cornetcies on the northern frontier the scarcity of suitable building timber was still a huge problem. To fetch it in the Cape was out of the question, and timber was scarce in these areas as well. It was especially difficult in the Roggeveld to obtain good timber. This was one of the reasons why the houses in this region were so poor and small even after it had long been inhabited.[53] Lime was also scarce in the interior. It could not be burned everywhere and there was not always money to buy it. As a result the smart, lime-plastered houses that were generally to be found in the Cape environs gradually became less frequent as a person traveled north from the coast.[54]

A shortage of good building materials, however, was not always the principal reason for poor housing. Travelers often observed to their surprise that even in regions where good timber was easily obtainable, pioneer houses were just as poor and simple. In Camde-boo, which lay only about four or five day's journey from the for-ests, pioneer dwellings were no better than those of farmers in the

Sneeuberg. And the houses in Outeniqualand and along the Little
Fish River, which was close to the forests, were just as bad.[55]

Travelers who did not always have a good understanding of
conditions in the Cape Colony were inclined at times to place the
blame for poor housing in the country areas on the colonists' "laziness." Paterson remarked, for example,

> The following day I visited one of the Dutch Boors who had
> resided in that part of the Country [along the Sundays River]
> for many years. This man was possessed of numerous herds of
> cattle: but had no corn and scarsely a house to live in, though
> the place was favourable to both. But the generality of these
> people are of so indolent a disposition, that they seldom trouble
> themselves either to build houses or to cultivate the ground.
> Those of them who chuse to be industrious, and to make the
> most of their advantages, are enabled to live in a very comfort-
> able manner.[56]

Laziness is as old as mankind and appears wherever there are
people. Thus we can assume that, in fact, laziness would have prevented the construction of decent houses in some cases in the frontier districts. But one must be very careful not to exaggerate,
because the pioneers' apparent indifference toward their living conditions can easily be attributed to other causes. These pioneers on
the outposts of civilization were often the descendants of people
who were *compelled* by circumstances to be content with what provided for their primary material needs alone. The pioneers were
satisfied therefore with the low living standard that they had received as a legacy from the previous generation. And because pioneer children generally carried on the migration into unknown
territories, there was a tendency throughout the pioneer period in
the always shifting frontier zones to maintain a low standard of
civilized life. In addition, models that might have stimulated change
or improvement were usually lacking in areas of newer settlement.
Besides, the pioneers were isolated as a group from the rest of the
world and few opportunities existed for them to come into contact
with progressive farmers in regions of older settlement. There was
therefore little to encourage them to strive for a higher standard
of living.[57]

The insecurity of living along the borders, because of the natives, was another factor that retarded the proper development of farms. Farmers on the northern frontier could not live at peace on their farms after 1770. They were repeatedly driven from their land by the Bushmen. And those who were never really driven off always had to reckon with the possibility that it could happen.[58] On the eastern frontier the Xhosa presence forced the farmers to live in a state of continual uncertainty. They never knew when the Xhosa would again attack the colony and burn their farms, but they could never lose sight of the possibility that they might be forced to flee. Collins submitted the following report in 1809, for example, regarding Swagershoek, one of the most productive parts of the colony:

> Its proximity to Cafferland, which has subjected this and the neighbouring districts to so many misfortunes, prevents the building of good houses; but some of the habitations, particularly those situated on the Little Fish River, are provided with gardens well stocked with fruits and vegetables.[59]

Undoubtedly the poor housing in the country areas was in large measure a direct result of the high degree of mobility found at all levels of the pioneer population.[60] Most farmers during this period probably owned fixed family dwellings, but they seldom remained settled on a specific spot. Even the owner of a loan farm clearly exhibited a tendency to change residences every few years. This made the farmers unwilling to erect fine or solid dwellings on their farms. Such farm construction was attended with too much risk as long as they had to allow for the possibility that they would migrate again. Moreover, people are not so concerned about the comfort of their homes if they know it is only a temporary dwelling. Apart from the fact that the old pioneers frequently changed their family dwellings, there were also very few farmers who did not make a regular migration with their livestock every year. This temporary migration was undertaken because a change of "pasture and climate" was good for the livestock, in order to find better water and grazing land for the livestock, or in order to avoid cattle sickness. In those parts on higher ground—like the Bokkeveld, the Roggeveld, the Nieuweveld, and the Sneeuberg—it became too cold during the winter, which was exactly at lambing season, for the livestock.

In addition, the scarcity of firewood made the winter cold and unpleasant for the people as well. As a result the inhabitants of these regions moved down with their families and livestock for a few months each winter to the surrounding plains. There the climate was milder and the grazing land lusher during the winter. In other regions—like the Hantam, the Onder-Bokkeveld, and the field-cornetcies to the north of the Sneeuberg—a shortage of pasture and water during the summer again led to regular seasonal migrations. In the Roggeveld a more or less double migration occurred yearly on a regular basis. During the winter months farmers were compelled by the cold to move down to the warmer Karoo; in the summer, more often than not drought compelled them to migrate to the Sak and Riet rivers.

Where seasonal migrations occurred regularly, it is understandable that a farmer would not go to any great trouble to build a good house on his farm, since it was uninhabited for much of the year. Paravicini di Capelli made the following observations in his travel journal after his party visited the farm of Commandant Johannes van der Walt next to the Seekoei River,

> It is peculiar to see such a substantial farmer, possessing numerous cattle and sheep, living then with his family in one wretched reed hut and having to make do under the most cramped conditions; nevertheless these people are so accustomed to living this trekking life (which they must do annually to graze their livestock), that they find no deprivation or inconvenience in it.[61]

In 1805 General Janssens also pointed out the detrimental consequences of the unavoidable seasonal migrations:

> There were many farms that had cattle fodder for only three or four months of the year—on other farms the herds die away in certain seasons—and other farms again, though barren and poor, are nevertheless usable for cattle for a few months, with the fortunate advantage that those cattle, which have grazed there for those months, can survive on the otherwise unhealthy pasture for the remaining part of the year without becoming sick. Thus some farmers must roam about the country out of *necessity*, and those who are accustomed to it do it in the end

partly out of choice. Learned men, full of self-conceit as they journey through the country, ascribe all the imperfections to the farmers' stupidity and base qualities. Yet there is much to challenge in such judgments. One point alone serves to underscore the stupidity of these scholars—how will men be found who will build handsome houses, plant fine gardens and tree orchards, sow the fields, etcetera, in a place where the people live but for a short time and where for much of the year they have to be away, during which time everything would fall into disrepair, be destroyed, or stolen.[62]

Apart from periodic migrations, the population of every new frontier zone was generally very mobile during the first few years. Large pieces of land lay uninhabited and this offered the pioneers the opportunity to wander around with their livestock in pursuit of water and rich grazing land. The hunt for game also contributed much to keeping the frontier population in a state of constant movement. Under these conditions the population of a newly settled area did not maintain fixed dwelling places initially, and consequently they did not erect permanent abodes for the first few years. The pioneers simply lived in their wagon tents or, if perhaps they lingered a while at a fountain or a water pan, they built reed huts, or mat or straw houses. Such temporary living quarters could be constructed quickly and left behind without serious loss when the site was abandoned.[63]

Farmers who settled on permanent dwelling sites in new pioneer areas also often had to make do initially with huts that could be erected in haste. They were often poor and therefore could not afford to build good houses immediately. Accordingly, in areas of more recent settlement, the wattle-and-daub hut was generally the forerunner of the "walled house." The latter appeared only after conditions were more settled, farmers enjoyed greater prosperity, or moneyed farmers from areas of older settlement had begun to move in.

When Masson visited the fertile Langkloof in 1773 there were still no walled houses. "It contains about seven or eight places, which are twelve to twenty miles distant from each other; the houses are very mean, without walls, consisting only of poles stuck in the ground, meeting at the top, and thatched over with reeds."[64] About a year before, Sparrman had visited the Krom River. "A farmer has lately settled here, and begun to cultivate a portion of land," he reported in his travel journal. "Up till now he has nothing in which to live save one hut made of leaves and straw."[65] In 1834 Pringle described the hut of a pioneer in the Zuurberg as "a wigwam constructed of a few poles and reeds."[66] A few years after a pioneer area received its first permanent inhabitants, however, the wattle-and-daub huts began to make room for the normal type of country farmhouse.[67] Housing in a specific region was therefore also directly dependent on the stage of economic development that that area had reached.

In most of the Cape Colony the nomadic type of stock farmer who maintained no permanent family dwelling disappeared fairly quickly, or at least became more rare, as the population density of the area grew. Livestock migrations still occurred until very recently in parts of the northwestern Cape, where nomadic grazing of the land continued to be dictated by the inhospitable nature of the country. Throughout the nineteenth century farmers on crown land migrated around the countryside without settling down in any one place. They made use of the land on the basis of the "grass license," which gave them the right to migrate wherever the rain fell. Thus until a relatively short time ago Bushmanland and Namaqualand, for example, barely supported a small nomadic population that constantly migrated from one pan or piece of green field to the next, always tracking the thundershowers that fell now here and then there.

In these circumstances it became the tradition to live in simple reed huts that could be easily erected and then taken down again, and which were light enough to transport from one place to another.[68] In 1929 the Poor Whites Commission still ran across people in Namaqualand who for the greater part of their lives had not lived in houses and whose children were all born in reed huts.[69] "In some areas of the northwestern Cape," stated the Reverend Albertyn, "there are still places today where people talk about walled houses as distinct from reed or wattle-and-daub huts because the latter are still so often to be found."[70]

In other areas of the northwestern Cape that were only colonized late in the nineteenth century and where the tradition of reed huts did not exist, the pioneers lived for years in tents. In districts such as Philipstown and Prieska many farmers during the last quarter of the nineteenth century still migrated around from one pan to the other, without setting up permanent dwellings. Among some unsorted documents in the Cape archives I accidently found in 1934 a petition that sheds interesting light on this subject. It was drawn up in 1888 by seventy farmers from the constituency of Prieska. Only eight of those farmers possessed less than 500 head of sheep and goats and eleven of them had more than 1,500. They all lived in tents on quitrent property, and their grievance was that they were scratched from the voting list because they did not occupy houses.

The Farmers' Furniture

Among rich farmers in the interior one could now and then run across some pieces of respectable furniture, but generally the farmer's dwelling was fairly meagerly provided with articles of furniture.[71] What there was, was brought out on a wagon from the Cape or was made on the farm itself. In the front room there would usually be at least one or two tables and a few chairs and benches, with or without backs, the seats of which were either covered with leather or matted with interwoven throngs. The less privileged, however, often had to be content with a wagon chest instead of a table and to manage with a few simple field stools.[72] A sideboard was almost never found in the front room. Recesses in the walls usually took their place. Cleanliness was not necessarily lacking, however, even in such poverty-stricken conditions.[73]

Tablecloths and napkins were everywhere not in fashion in the country areas. As a general rule, knives, forks, and spoons were fairly rare. In remote areas knives, which in the old days were usually "hernhutters," were fairly commonly used at the table. These knives were made by the people on the Hernhutter missionary stations. They were normally seven or eight inches long, had a hilt of four or five inches, and were carried in a sheath on a belt. It was generally assumed that a visitor or traveler would have his own knife, and only forks were set down next to the plate.[74] Now and then, however, one would see silver cutlery in the interior. Rich farmers sometimes also possessed earthenware dishes or fine porcelain but this was not very common. By the time porcelain reached the frontier districts, it was for the most part already broken or cracked. The "crockery" of the frontier farmers most of the time consisted of unbreakable things. Dishes were often of wood. Plates were of pewter—not tin, but pure pewter, which could be smelted and molded into other objects or mixed with lead in order to make the bullets harder. Round earthenware bowls were used instead of cups and saucers. Furthermore, by the time of the Great Trek the farmers had large copper jugs that cost a pound sterling each and lasted forever.[75]

In the bedrooms of some farmhouses there were old-fashioned bedsteads resplendent with high canopies and bed-curtains, and neatly fitted with soft feather mattresses, woolen blankets and stitched bedspreads. More commonly, however, the bedstead was much simpler. Usually it consisted of a square wooden framework that rested on four legs, each about two feet high. Most of the time the bottom was of leather thongs, although woven rattan was also employed occasionally. Skin blankets were very popular. Sometimes there were extra beds and bedding for travelers, but it was also common for guests to have to sleep on mats on the ground, with their jackets over their bodies and their saddles under their heads.[76]

Nutrition in the Outlying Districts

In the olden days in the interior meat was the staple dish. Bread and vegetables were usually scarce, particularly in areas where pioneers had just arrived. Even where gardening and agriculture were possible, cultivation was out of the question as long as the pioneers were still unsettled and moved round in search of game and grazing land. Besides, the first arrivals in a new pioneer area were often poor. They frequently had migrated to the frontier with only their wagons, rifles, and a handful of livestock, and sometimes it was a long while before they could afford a plow.[77] More often than not they had no money to buy grain and, besides, there was nowhere in the vicinity that they could get it. Consequently, the livestock farmer-hunter on the borders of the colony in the eighteenth and nineteenth centuries—just like the migrant farmers in Kolbe's day—often ate dried game in place of bread with their fat mutton.[78] Bread became a part of the farmers' daily diet again only a number of years after they first entered a new area and living conditions were somewhat stabilized.[79] Nearly every pioneer area went through this process.[80]

Consequently, a frontier traveler could go for days sometimes without having any bread or vegetables. Swellengrebel's daily log of his journey in 1776 bears witness to this. From Willem Prinsloo's farm near Bruintjieshoogte the return trip passed all along the eastern border and then along the coast to the Cape. On the farm of P. Buys alongside the Swartkops River the party saw cultivated land again for the first time since leaving the Prinsloo farm. On 18 November the party happily received from one Potgieter bread, butter, and vegetables, after having to manage from the beginning of the month without these luxury items.

A few days later (30 November) the following entry was made in the diary:

In the plain between the hills of the Lange Kloof and the Kammanassie mountains, as well as along the Olifants River's mountain range, till where the Kauwkwa turns toward the sea and the mountain range ends, are located several farms, at the most 3 to 4 hours from each other, where there are still men, as has been testified and I myself experienced at the warm spring (where men have had no bread since April), who have no bread most of the time and have to eat flesh with flesh. The prolonged droughts, as a result of which the land cannot be plowed and sowed early, makes grain agriculture very difficult, especially here in these inland parts. Yet I think at the same time there is something wrong here with the industriousness, because although each farm is not situated on a great river, at least it lies on a little river or brook, and there is thus quite enough damp terrain as is necessary for a garden. People here know little of vegetables. Even if the land here about in the Lange Kloof was very infertile, there is an abundance of water, and men could think about living more respectably. But due to the prohibitively high expenses attached to the transport of grain, men concentrate only on stock breeding and sink indifferently into the lazy life of a pastoralist. The farther they move away from the major city, the worse the danger becomes for most colonists.[81]

Undoubtedly certain types of pioneers were relatively uninterested in land cultivation, but there has been considerable exaggeration in this respect in the past. Barrow states, for example, that farmers seldom ate bread,[82] and regarding Graaff-Reinet he relates,

> The boors [sic] of this district are entirely graziers; few attempting to put a plough or spade in the ground, except in Zwartkops Bay, or in some parts of the Sneeuwberg, preferring a life of complete indolence and a diet of animal food to the comfort of procuring a supply of daily bread, and a few vegetables, by a very trifling degree of exertion. In Sneeuwberg, indeed, the depredations of the locusts are discouraging to the cultivator, as the odds are great he reaps nothing, while this devouring insect remains in the country.[83]

This picture of frontier life could not possibly be altogether accurate, because Governor Janssens made an extended journey through the colony shortly after Barrow and he concluded that

> in the areas where Barrow said the inhabitants eat no bread, through laziness and other causes, I came across bread everywhere, and it was especially good. And it was not any special bread for the governor, the inhabitants consumed it all about. The commissioner general, traveling later and being in sections of the colony which were not visited by me, also found bread being eaten wherever there were inhabitants.[84]

But Janssens made the opposite kind of mistake. His account creates an impression that is too rosy, because bread actually was scarce at times in the country. The truth appears to be that the old farmers generally exerted themselves to lay out gardens and agricultural plots wherever they had been settled for a while, but they were not always very successful in their undertakings. This is evident from a letter written in 1810 by Landdrost Stockenstrom of Graaff-Reinet, who undoubtedly was well informed:

> None of the Inhabitants, who have places which will admit of it, neglect growing as much corn and barley, or cultivating as many vines and fruit trees, as they require, not only for their own use, but also for supplying the wants of those neighbours,

whose places either do not contain land which will grow such produce, or do not possess the advantage of a command of water; these persons, however, are even often disappointed in their crops, owing either to the rains falling late or to the whole crop being destroyed by locusts, which I have seen happen five years successively.[85]

Stockenstrom's report is supported by a declaration of C. H. Olivier, a butcher from Graaff-Reinet, who through his trade in cattle became familiar with the district. In 1826 he was asked on what the farmers lived. His answer was,

Meat and milk. Many of them have not tasted bread for six years. They used to get meal from the corn farmers, but these now rather prefer to sell their grain to the troops on the Frontier for money, than to barter it for cattle. They try to grow corn every year, but unless there is rain in January and February, the crop is burnt up, and the seasons have been very bad.[86]

In many other parts of the colony things were not much better than in Graaff-Reinet. Sowing seed in the sheep farming districts was often altogether impossible because of inadequate rainfall or an insufficient supply of irrigation water and arable land. Farmers in such zones had to make do then without bread if they had no money with which to purchase flour.

Where farmers could not lay out gardens, they seldom ate vegetables. Howison cites the case, for example, of a farmer who lived alongside the Leeuw River in the Karoo, Beaufort district. When he visited him, the farmer had had no rain for the previous four years. The river had become brackish and the water was undrinkable. He had tried to plant a garden in the saline soil but to no avail. His livestock had died from the drought and he could not shoot wild game because he was out of ammunition. He had not eaten any bread for weeks. But this was not an unusual case. Hundreds of farmers in the interior, as Howison assures us, struggled with such problems and were forced to endure similar hardships.[87]

While Lichtenstein was busy giving vaccinations against smallpox, he remained a long time at the home of his old friend van der Westhuizen, at the foot of the Roggeveldsberg. He later explained,

During this period, however, I was deprived of many things, and lived even more meagerly than in the latter half of my recent journey. My host and all his neighbors subsisted almost entirely on the flesh of mutton, so that around here scarcely any other foodstuffs were to be found. Two or three sheep were slaughtered daily; from these the cooked entrails and feet were served for breakfast, (a dish well-known throughout the entire colony as *spleen and trotters*), together with the roasted fat of the tail. The noon meal generally consists of a strong soup and roasted mutton; and in the evening the boiled ribs were our supper. Everything was very tasty and cleanly prepared but without bread, and even salt (of which there was little to be had). It truly required a healthy appetite to be able to live on such fare for many weeks in succession. My sincere host, to whom I complained that the absence of bread was the most difficult for me, spared no pains in seeking to obtain some from his neighbors; but a hat full of barley was all he brought home. With this the good wife cooked me a suitable dish, which no one else was allowed to taste.[88]

In other outlying districts where the rainfall was not so uncertain, things went better for the farmers. In the Onder-Bokkeveld, for example, the majority of farmers harvested their own grain. Many could even barter or sell surplus grain to the farmers of the Hantam, the Roggeveld, and Namaqualand—where less grain was harvested. In the latter districts the crop was also good when it had rained, but it often happened that the farmers did not recover their sowing seed.[89]

Thus, an exceedingly large quantity of meat was consumed under these conditions in the livestock districts on the northern frontier. It was the only sort of food of which there was never a shortage. Every livestock herdsman normally received a slaughter animal per week. In large households where there were many children and servants, three or four sheep, weighing from thirty-six to forty pounds each, were slaughtered daily. Lichtenstein and the Commission for Stock Raising and Agriculture assure us of this.[90]

Although many farmers in the interior sometimes lived for weeks and months on nothing besides meat, other farmers who owned gardens and agricultural plots could eat in abundance. On their tables the visitor could usually indulge not only in all sorts of meat dishes, but also bread and various types of vegetables, salads, and fruit.[91] And on special occasions the tables were literally piled high with an abundance of everything the garden, farm, and pasture had produced. We are assured by one trustworthy traveler that at times a person could hardly count the dishes that appeared with each course on the table and there was also no end to the number of courses. Mealtime usually began with a hearty meat soup, in the making of which red pepper and all kinds of spices were not spared. As hors d'oeuvres, curry and rice or calabash stew were often served. The latter dish was typically South African, made with pumpkin, finely chopped onion, salted sea fish, and cayenne pepper, all mixed together and cooked. Even Northern Europeans, who normally did not think much of the highly seasoned Cape food, found the calabash stew a tasty dish.

Following the hors d'oeuvres came cooked beef or sea fish with pickles and condiments, and after that the barbecued meat. This usually consisted of a sucking pig, a turkey, or some game. Served with the meat dishes were preserved fruit, of which there were sometimes seven or eight types, each in a separate dish. Vegetables were seldom prepared with butter. Tail fat from mutton was used most of the time in place of butter, but it was so tastefully prepared that even the most finicky gourmet found it delicious. Different sorts of pies completed the warm courses, whereupon an excellent dessert followed, most of the time consisting only of fruit. Just to gaze on all the varieties of fruit was already a delight. Many Europeans, who were accustomed to much less, regarded the fragrant and multicolored pyramids in the fruit bowls of the simple farmers with an envious gaze.

Male slaves, or Khoikhoi maids in poorer areas, normally performed the kitchen work, but the housewife kept a close watch over all the activities so there was no lack of tidiness. During mealtime many of the slaves waited at table, while others stood behind the guests and kept away the flies with ostrich feather fans. Frequently there was music. Before and after eating the guests had the opportunity to wash.[92]

Tea was the favorite drink in the country areas during the eighteenth century and the first years of the nineteenth century. The constantly filled teapot on the little table in the drawing room was one of the first things that struck the traveler. Any time of the day one could count on at least a cup of tea. In regions where the water was brackish, tea was drunk from morning till evening. The tea was generally weak and most of the time consumed without sugar and milk. It sometimes happened, however, that a small tin of sugar was passed around, so that everyone could take a small lump to hold in his mouth while he drank the tea. Milk was not used very much. If the cows dried up or were sent away from the farm to find better grazing land elsewhere, the farmers had to make do without milk altogether. Milk was abundant in regions where cattle farming was predominant, but most of it was used for butter production. Beer was nowhere to be found in the colony, except in Cape Town. Wine was not drunk in the interior, but now and then a small vat of brandy was obtained from an itinerant merchant or brought out from the Cape.[93]

Fashions in the Country Areas

In the olden days the farmers also adopted their fashions from Europe.[94] Even in their isolation they never developed a truly national clothing style. European fashions were usually a year or so old by the time they reached Cape Town. From there they spread even more slowly inland. The small villages followed Cape Town relatively quickly, but it took a very long time before a new style reached the homesteads in the interior. The fashions of the frontier farmers were generally at least ten years out of date—if we take Cape Town as the norm.

By the time of the Great Trek one could generally distinguish two sorts of attire for men in the country areas. The Doppers wore short jackets and flaptrousers, which were buttoned at the hips. Most of the time these trousers had drawstrings, which prevented them from slipping down. At times there was also a belt fastened around the stomach. Neither the belt nor the fastener could be trusted, however. The trousers frequently sagged down on the hips, and since the jackets were so short, there was then a piece of shirt that stuck out between the trousers and the jacket. The other farmers never had any shirt showing at the back. They were prudent enough to wear suspenders and, besides, their jackets were longer than those of the Doppers. Trousers were often of different lengths, but they usually reached at least to the ankle, and some of them were worn to the top of the shoe. Leather trousers, however, were usually on the short side. The bottoms of the legs became wet too easily in the grass and then the leather naturally became stiff and hard as it dried again.

Men's clothing was usually made of "duffel" material that was woolly in appearance and gray or brown in color. Sometimes moleskin, buckskin, or nankeen was used. Blue with a stripe was very popular, although the Doppers generally chose brown. For their "Sunday best," which was only worn on Sundays or for special occasions, a better material was used. These clothes were usually made of smooth or corduroy velvet—black, green, brown, or dark yellow in color. Only rich farmers could afford a cloth coat and vest with cashmere trousers. During the week an ordinary checked or blue woolen shirt was worn, but on Sunday it was exchanged for a shirt of fine linen.

Various sorts of hats were in fashion. The majority of farmers chose wide-brim hats that could protect their faces and necks against the stinging rays of the sun. Narrow-brim hats of otterskin with high crowns were also seen now and then. By the time of the Great Trek children frequently wore straw hats that were personally woven by their mothers. Andries Hendrik Potgieter, the famous Voortrekker leader, also owned a straw hat, with a wide brim and a green lining,[95] and at times Piet Retief wore a woolen hat with a small brim and a rather high crown.[96] Such hats were usually made by the farmers themselves and easily lasted for five years. Gerrit Maritz migrated in a top hat of gray wool, but this was surely an exception.[97]

By the time of the Great Trek women's fashions displayed greater variation than men's styles. Nevertheless, dresses usually had collars around the throat and wide sleeves, which were fastened with a button at the wrist. In addition, the dresses were worn long, and they frequently had a multitude of tucks from the hem to approximately the height of the calf. Shawls of wool or silk were often seen. Also by the time of the Great Trek ladies generally no longer went walking without stockings, which fifty years earlier was common practice. Men, however, did not always wear socks and children hardly ever. Gloves were a rarity, but ladies often used mittens. Both sexes sometimes purchased ready-made shoes in the store or from an itinerant merchant, but the majority worn home-made rawhide shoes. In fact, in the early days handmade leather thong shoes were more highly regarded than store bought shoes.

Ladies usually parted their hair in the middle and the hair hung down in two braided strands behind their back. Now and then it was pinned together with a comb of tortoise shell. At times women and girls wore bonnets and used umbrellas to shade their faces against the rays of the sun, but the typical headcovering for women was the famous "kappie" or sunbonnet. The sunbonnet fitted flat over the head like a wagon tent and concealed the face in the deep recess while the frills reached down to the shoulders, protecting the neck and throat against the sun and wind. This was the principal merit of the sunbonnet, because the women then were very proud of their facial color, and did not want to be sunburned "like kaffirs."

Making a sunbonnet was an art in itself because it had to be completely stitched by hand. It also took a long time to make. It was prepared with the greatest care and attentiveness, and all kinds of patterns were often embroidered on the crown—little hearts, diamonds, or circles, or a combination of all sorts of forms—while the frills were decorated with pleats and hems. A sunbonnet was tied together with two strings under the chin and would thus not blow away. Usually sunbonnets were made of fine white linen, but dark merino was also used now and then. Sunbonnets were easily washed, but the ironing was more complicated. In order to give it the necessary rounding over the head, a sunbonnet had to be ironed over a block.

Women's clothing in the old days was mostly of chintz, merino, alpaca, check, and cambric. Silk dresses were not a rarity, but were worn only for special occasions. Women's undergarments were usually made from linen, red or white flannel, check, or baize.[98]

Clothing fabrics were commonly purchased by the roll if the wagon went to the Cape or if an itinerant merchant appeared.[99] Woven fabric, however, was expensive and there was not always an itinerant merchant or a store near at hand. Therefore the farmers tried to substitute store-bought clothing fabrics as much as possible with the leather that they themselves prepared. Leather trousers especially were commonly used throughout the interior. And it was not only the farmers who wore leather trousers in those days. Every second year the Company bought 1,800 sheepskins "to be able to provide the soldiers with new leather trousers to replace their old used clothing."[100] Other articles of clothing, like overcoats for men and overskirts and petticoats for women, were also frequently made from leather. For these garments they used thoroughly prepared leather that was usually as soft as cloth. Unfortunately it could not easily be washed, but it could be cleaned with orange juice and other similar substances. Dressed calfskin was sometimes used for vests with the hair side to the outside.[101]

In the course of the nineteenth century leather was replaced more and more by factory-made fabrics, but skin clothing was still not uncommon, even among well-established farmers, at the time of the Great Trek. Frequently children, slaves, and Khoikhoi in the outlying districts wore nothing else.[102] And the poor—who were also with us in the early days, although then they could more easily maintain their independence, and their poverty did not carry with it so much social inferiority as today—could usually not afford woven clothing fabrics. They had to manage as well as they could with skins.[103]

NOTES

1. A. Sparrman, *Reize naar de Kaap de Goede Hoop, de Landen van den Zuid Pool, en rondom de Waereld* (Leiden, 1786), I:151.

2. T 53, Letters Rec., Petition from the Koue-Bokkeveld, 22 December 1818. [Trans. note: There are three "T" documents in *Trekboer* but vdM does not list them in the bibliography. At the end of his bibliography in *Noordwaartse Beweging* (p. 400), he does give his code of abbreviations and T is for the Tulbagh Archives. Of the other two documents in this work, the first, in chap. II, note 26, is now in CO 2614, and the third, in note 9 below, is now in 1/WOC 1/6. Searches were made in the records for Tulbagh as well as in the relevant locations in the 1/WOC and CO records but this petition was not found.]

3. Swellengrebel Family Archives, Journal of Swellengrebel's Journey, 1776; C 1006, Enclosure, Petition from the Bokkeveld 18 September 1792, p. 116; Commission for Stock Raising and Agriculture, dated 20 November 1805, in G. M. Theal, *Belangrijke Historische Dokumenten over Zuid-Afrika* (Cape Town, 1896, 1911), III:345; H. Lichtenstein, *Reisen im*

südlichen Africa in den Jahren 1803, 1804, 1805 und 1806 (Berlin, 1811–1812), I:207.

4. Janssens to de Mist, 9 April 1803, in Theal, *Belangrijke Historische Dokumenten*, III:208. Cf. C 123, Resolutions, 23 March 1745, p. 121.

5. Augusta Uitenhage de Mist, "Dagverhaal van een Reis naar de Kaap de Goede Hoop en in de Binnenlanden van Afrika door Jonkvr. Augusta Uitenhage de Mist in 1802 en 1803," in *Penélopé*, VIII:83–84; T. Philipps, *Scenes and Occurrences in Albany and Cafferland, South Africa* (London, 1827), p. 14; "Four Months in the Cape," in *Chambers' Miscellany of Useful and Entertaining Tracts*, XX(173):29; Lichtenstein, *Reisen*, I:36.

6. Sparrman, *Reize*, I:153 and 161; Lichtenstein, *Reisen*, I:35.

7. C 156, Resolutions, 1 December 1778, p. 389; C 186, Resolutions, 23 March 1790, p. 508; C 199, Resolutions, 20 December 1791, pp. 659–60; Sparrman, *Reize*, I:271, 301.

8. W. J. Burchell, *Travels in the Interior of Southern Africa* (London, 1822), I:52.

9. Testimony of C. H. Oliver, 8 December 1826, in G. M. Theal, *Records of the Cape Colony, 1793–1831* (Cape Town, 1897–1905), XXIX:478; 1/WOC 1/6, Resolutions (Worcester) 5 February 1827; E. C. Godée Molsbergen, *Reizen in Zuid-Afrika* (The Hague, 1932), II:266; H. Lichtenstein, *Reisen*, II:578; J. F. W. Grosskopf, *Plattelandsverarming en Plaasverlating* (Stellenbosch, 1932), pp. 32–33.

10. Burchell, *Travels*, I:201–2; P. B. Borcherds, *An Auto-Biographical Memoir* (Cape Town, 1861), p. 55–57; O. F. Mentzel, *Vollständige und Zuverläszige . . . Beschreibung . . . des Guten Hoffnung* (Glogau, 1785, 1787), II:351; Lichtenstein, *Reisen*, I:33, 38; Sparrman, *Reize*, I:282; C 171, Resolutions, 19 April 1786, p. 428.

11. AR, Accession 157 (Nederburg), van Ryneveld to Janssens, 4 September 1804.

12. A. J. H. van der Walt, *Die Ausdehnung der Kolonie am Kap der Guten Hoffnung, 1700–1779* (Berlin, 1928), pp. 28–40.

13. Lichtenstein, *Reisen*, I:72; I:85–86.

14. VC 96, Journal of J. W. Cloppenburg, 1768, p. 7.

15. A. L. Geyer, *Das Wirtschaftliche System der Niederländischen Ostindischen Kompanie am Kap der Guten Hoffnung* (Munich and Berlin, 1923), pp. 70–74.

16. Sparrman, *Reize*, I:281; Lichtenstein, *Reisen*, II:9, 112.

17. PSB, Mss. Germ. Fol. 879, J. A. de Mist, Reisverhaal, p. 227; Lichtenstein, *Reisen*, I:25; Barrow, *Travels*, II:331; A. U. de Mist, "Dagverhaal," p. 83.

18. C 172, Resolutions, 4 August 1786, p. 917.

19. AR, Accession 40/45a, Petition dated 9 October 1779, p. 34; Barrow, *Travels*, II:331; Swellengrebel Family Archives, Journal of Swellengrebel's Journey, 1776, p. 23.

20. PSB, Mss. Germ. Fol. 879, J. A. de Mist, Reisverhaal, p. 277; A. U. de Mist, "Dagverhaal," p. 83; Lichtenstein, *Reisen*, I:33.

21. "Remembrances of L. C. de Klerk," in G. S. Preller, *Voortrekkermense* (Cape Town, 1920–1925), I:216.

22. Lichtenstein, *Reisen*, I:75, 117; Barrow, *Travels*, II:402; H. Raikes, *Memoirs of the Life and Services of Vice-Admiral Sir Jahleel Brenton, Baronet, K.C.B.* (London, 1846), p. 555; J. B. N. Theunissen, *Aantekeningen van een reis door de Binnenlanden van Zuid-Afrika van Port Elizabeth naar de Kaapstad gedaan in 1823* (Oostende, 1824), p. 88: J. W. D. Moodie, *Ten Years in Southern Africa* (London, 1835), p. 62; H. H. Methuen, *Life in the Wilderness; or Wanderings in South Africa* (London, 1848), p. 31; "Remembrances of L.C. de Klerk," in Preller, *Voortrekkermense*, I:249ff.; Commissioners of Inquiry: Report II, Upon the Finances at the Cape of Good Hope, 6 September 1826 (Cape Town, 1827), p. 87.

23. C 1267, Petitions, Reports, etc., Petition of the Military officers of Camdeboo, 15 March 1777; 1/GR 1/1, Resolutions, 13 November 1786, pp. 1–6; C 563, Letters Rec., Woeke to Governor, 11 July 1786; C 1300, Petitions, Reports, etc., Petition from Camdeboo, 1 February 1788, new pp. 223–37, and Woeke to Governor, 5 January 1789, new pp. 214–18; C 1007, Enclosure, C. van der Merwe et al. to Landdros, 10 October 1790, pp. 139–42; 1/GR 1/1, Resolutions, 21 December 1791, pp. 171–73; Macartney, Instructions to the Landdrost of Graaff-Reinet, 20 June 1797, in Theal, *Records*, II:95, 101.

24. C 113, Resolutions, 19 January 1740, p. 49; C 1019, Enclosures, Report from G. Goetz to Governor Rhenius, 11 December 1792, pp. 393–446.

25. Kaapsche Geschillen (Cape Disputes) (The Hague, 1785), J. van Plettenberg to Lords XVII, 20 March 1781, III:12. [Trans. note: One copy of this rare volume is located in the South African Library, SAL A.Z.968.7025 Gen, under the binder's title, *Old Cape History*.]

26. PSB, Mss. Germ. Fol. 879, de Mist, General Observations (1803), p. 232.

27. Sparrman, *Reize*, II:594; Lichtenstein, *Reisen*, I:266.

28. Mentzel, *Beschreibung*, II:168.

29. Swellengrebel Family Archives, Journal of Swellengrebel's Journey, 1776, p. 23.

30. Ibid., entry for 1 November 1776, pp. 23, 27, 49.

31. CO 2695, Letters Rec. from Graaff-Reinet, no. 153, Graaff-

Reinet Landdrost and Heemrade to Colonial Secretary, 31 December 1827.

32. C. de Jong, *Reizen naar de Kaap de Goede Hoop* (Harlem, 1802–1803), II:139.

33. In *Boer en Barbaar* by J. H. Malan the reader will find a chapter discussing the Cape architecture of the eighteenth century. [Trans. note: vdM did not indicate in his bibliography the edition to which he is referring. It is assumed he intended the reader to look at chap. 5, for which the material in the two volumes is substantially the same. J. H. Malan, *Boer en Barbaar; of, Die Lotgevalle van die Voortrekkers viral tussen die jare 1835–1840* (Potchefstroom, 1913), chap. 5, pp. 123–33, or *Boer en Barbaar; of, Die Geskiedenis van die Voortrekkers tussen die jare 1835–1840, en verder, van die Kaffernasies met wie hulle in aanraking gekom het* (2d expanded ed., Bloemfontein, 1918), chap. 5, 90–95.]

34. PSB, Mss. Germ. Fol. 879, J. A. de Mist, Reisverhaal, pp. 330, 352–53.

35. Cf. Lichtenstein, *Reisen*, I:76.

36. Swellengrebel Family Archives, Journal of Swellengrebel's Journey, 1776, pp. 11, 72, 88.

37. Journal of van Plettenberg's Journey, in Theal, *Belangrijke Historische Dokumenten*, I:29.

38. Lichtenstein, *Reisen*, I:167; Godée Molsbergen, *Reizen*, II:182.

39. PSB, Mss. Germ. Fol. 879, J. A. de Mist, Reiverhaal, p. 358; AR, Accession 1900, no. XX, Journal of Governor Janssens' Journey, 1803; Burchell, *Travels*, I:198, 237; C. J. F. Bunbury, *Journal of a Residence at the Cape of Good Hope, with Excursions into the Interior* (London, 1848), p. 181; Lichtenstein, *Reisen*, I:168.

40. F. Masson, "An Account of Three Journeys from Cape Town into the Southern Parts of Africa," in *Philosophical Transactions of the Royal Society* (London, 1776), p. 299; Sparrman, *Reize*, I:149, 322; Lichtenstein, *Reisen*, I:168; Barrow, *Travels*, II:401; Burchell, *Travels*, I:237; G. Thompson, *Travels and Adventures in Southern Africa* (London, 1827), p. 82; A. Steedman, *Wanderings and Adventures in the Interior of Southern Africa* (London, 1835), I:146.

41. AR, Accessions 1914 (Van Plettenberg Collection no. 28), O. G. de Wet, Daily account of a journey, pp. 32–33.

42. Swellengrebel Family Archive, Journal of Swellengrebel's Journey, 1776, p. 27; cf. also Barrow, *Travels*, I:77, and Masson, "Account," p. 28. Le Vaillant describes the houses of the frontier farmers: "They are merely a barn, consisting of a single room, without any division" (*New Travels into the Interior Parts of Africa in the Years 1783, 1784 and 1785* [London, 1786], I:54).

43. Colonial Library, The Hague, Paravicini di Capelli, *Reizen in de Binnenlanden van Zuid-Afrika* (1803). [Trans. note: vdM does not cite page numbers for any of the Paravicini di Capelli notes. I have added the appropriate page number from the van Riebeeck Society edition. VRS, entry for 12 July 1803, series I, vol. 46, p. 162.]

44. Barrow, *Travels*, II:85.

45. Lichtenstein, *Reisen*, I:169; A. U. de Mist, "Dagverhaal," p. 95; Burchell, *Travels*, I:237.

46. Colonial Library, The Hague, Paravicini di Capelli, *Reizen* [V.R.S. entry for 29 April 1803, series I, vol. 46, p. 48.]; Ann Elizabeth Nightingale, *Gleanings from the South, East and West* (London, 1843), p. 20; Barrow, *Travels*, I:84, II:401.

47. PSB, Mss. Germ. Fol. 879, J. A. de Mist, Reisverhaal, pp. 353, 355; Bunbury, *Journal*, p. 181; T. Pringle, *African Sketches* (London, 1834), p. 176; Thompson, *Travels*, p. 82; Lichtenstein, *Reisen*, I:169; Thunberg, *Travels*, I:256; Sparrman, *Reize*, I:149.

48. PSB, Mss. Germ. Fol. 879, J. A. de Mist, Reisverhaal, p. 357; Colonial Library, The Hague, Paravicini di Capelli, *Reizen* [V.R.S. entry for 12 July 1803, series I, vol. 46, p. 160.]; Godée Molsbergen, *Reizen*, II:182; A. U. de Mist, "Dagverhaal," p. 95; Barrow, *Travels*, I:77, 78; Thompson, *Travels*, p. 84; Pringle, *African Sketches*, p. 175; T. Smith, *South Africa Delineated* (London, 1850), p. 16.

49. PSB, Mss. Germ. Fol. 879, J. A. de Mist, Reisverhaal, p. 357; Burchell, *Travels*, I:198.

50. Pringle, *African Sketches*, p. 176; Burchell, *Travels*, I:237; Barrow, *Travels*, I:84.

51. Pringle, *African Sketches*, p. 176; Bunbury, *Journal*, p. 181.

52. PSB, Mss. Germ. Fol. 879, J. A. de Mist, Reisverhaal, p. 375; Lichtenstein, *Reisen*, I:135; Burchell, *Travels*, I:198.

53. Godée Molsbergen, *Reizen*, II:183; Lichtenstein, *Reisen*, I:167.

54. PSB, Mss. Germ. Fol. 879, J. A. de Mist, Reisverhaal, p. 350; Lichtenstein, *Reisen*, I:167.

55. Sparrman, *Reize*, I:322; Swellengrebel Family Archives, Account of Swellengrebel's Journey, 1776, pp. 27, 33; AR, Accession 1914 (van Plettenberg Collection no. 28), O. G. de Wet, Daily Account of a Journey, pp. 32, 35.

56. Paterson, *Narrative*, p. 84.

57. R. W. Wilcocks, *Die Armblanke* (Stellenbosch, 1932), p. 66.

58. Van der Merwe, *Noordwaartse Beweging*, pp. 7–12.

59. Collins, Journal of a Tour, 1809 (ms. in the Cape Archives).

60. The mobility of the pioneer population at the Cape will be discussed more fully in another work.

61. Colonial Library, The Hague, Paravicini di Capelli, *Reizen.* [V.R.S., entry for 23 July 1803, series I, vol. 46, p. 177.]

62. PSB, Mss. Germ. Quarto 867, Janssens, Petition, 30 January 1805, p. 60.

63. Masson, *Account,* p. 313; Thunberg, *Travels,* II:92; Barrow, *Travels,* II:117; Le Vaillant, *New Travels,* I:45; Lichtenstein, *Reisen,* I:35, II:108.

64. Masson, *Account,* II:290.

65. Sparrman, *Reize,* I:346.

66. Pringle, *African Sketches,* p. 202.

67. C 171, Resolutions, 19 April 1786, p. 427; Masson, *Account,* p. 290; Paterson, *Narrative,* p. 79; Lichtenstein, *Reisen,* I:339.

68. Barrow, *Travels,* I:386; Wilcocks, *Armblanke,* p. 66.

69. Grosskopf, *Plattelandsverarming,* p. 42; J. R. Albertyn, *Die Armblanke en die Maatskappy* (Stellenbosch, 1932), p. 6.

70. Albertyn, *Armblanke en die Maatskappy,* p. 66.

71. Swellengrebel Family Archives, Journal of Swellengrebel's Journey, 1776; B. Stout, *The Cape of Good Hope and its Dependencies* (London, 1820), p. 101; Thompson, *Travels,* II:82; Thunberg, *Travels,* I:256; Masson, *Account,* p. 288.

72. PSB, Mss. Germ. Fol. 879, J. A. de Mist, Reisverhaal, p. 455; Swellengrebel Family Archives, Journal of Swellengrebel's Journey, 1776; Howison, *European Colonies,* I:344 ; Burchell, *Travels,* I:198–99, 237; Barrow, *Travels,* I:77, II:420; A. U. de Mist, "Dagverhaal," p. 95; Lichtenstein, *Reisen,* I:119, 178, 371; R. Percival, *An Account of the Cape of Good Hope* (London, 1804), p. 216.

73. Lichtenstein, *Reisen,* I:168.

74. Sparrman, *Reize,* II:593; Thompson, *Travels,* II:117; Burchell, *Travels,* I:240; Barrow, *Travels,* II:420; "Remembrances of L. C. de Klerk," in Preller, *Voortrekkermense,* I:257.

75. Sparrman, *Reize,* II:593; Thompson, *Travels,* II:142; Percival, *Cape of Good Hope,* p. 216; "Remembrances of L.C. de Klerk," in Preller, *Voortrekkermense,* I:250, 251, 254.

76. PSB, Mss. Germ. Fol. 879, J. A. de Mist, Reisverhaal, p. 358; Lichtenstein, *Reisen,* I:119; Burchell, *Travels,* II:122–23; Barrow, *Travels,* II:401; Sparrman, *Reize,* I:149; Bunbury, *Journal,* p. 181; Pringle, *African Sketches,* p. 202; Mentzel, *Beschreibung,* II:174.

77. C 1283, Petitions, Reports, etc., Burgher Petition of 17 February 1784.

78. Mentzel, *Beschreibung,* II:13, 95.

79. C 1814, Letters Disp., Governor van der Graaff to Patria, 19 April 1786, pp. 527–28.

80. Mentzel, *Beschreibung*, II:19, 95, 341; Pringle, *African Sketches*, p. 177; Stout, *Cape of Good Hope and Its Dependencies*, p. 113; Allamand, Klockner, and Hop, *Nieuwste and Beknopste Beschrijving van de Kaap der Goede Hoop* (Amsterdam, 1778), p. 87.

81. Swellengrebel Family Archives, Journal of Swellengrebel's Journey, 1776.

82. Barrow, *Travels*, II:78–79; Howison, *European Colonies*, I:343.

83. Barrow, *Travels*, II:371.

84. AR, Col. Arch. 4375, Janssens to Asiatic Council, 8 January 1805, p. 275.

85. CO 2580, Letters Rec. from Graaff-Reinet, A. Stockenstrom to Colonial Secretary, 20 September 1810; cf. Commission for Stock Raising and Agriculture, dated 20 November 1805, in Theal, *Belangrijke Historische Dokumenten*, III:426; Paterson, *Narrative*, p. 79; Masson, *Account*, pp. 291, 293.

86. C. H. Oliver, 18 December 1826, in Theal, *Records*, XIX:478.

87. Howison, *European Colonies*, I:346.

88. Lichtenstein, *Reisen*, II:577–78.

89. Ibid., I:145, 167.

90. Commission for Stock Raising and Agriculture, dated 20 November 1805, in Theal, *Belangrijke Historische Dokumenten*, III:392; Lichtenstein, *Reisen*, I:134, 145, 602.

91. Borcherds, *Memoir*, p. 54; Burchell, *Travels*, I:220, 240, II:113; Lichtenstein, *Reisen*, I:294.

92. Lichtenstein, *Reisen*, II:135–40.

93. AR, Col. Arch. 4365, de Mist to Janssens, 22 November 1803; Commission for Stock Raising and Agriculture, dated 20 November 1805, in Theal, *Belangrijke Historische Dokumenten*, III:426; Godée Molsbergen, *Reizen*, II:183; Lichtenstein, *Reisen*, I:168; Howison, *European Colonies*, I:344; W. von Meyer, *Reizen in Süd-Afrika während der Jahre 1840 und 1841* (Hamburg, 1843), p. 165; A. U. de Mist, "Dagverhaal," p. 95.

94. "Remembrances of L. C. de Klerk," in Preller, *Voortrekkermense*, I:242, 243, 262.

95. Ibid., I:263.

96. Ibid., I:262.

97. Ibid., I:263; Borcherds, *Memoir*, p. 57.

98. "Remembrances of L. C. de Klerk," in Preller, *Voortrekkermense*, I:242ff. Cf. also Thunberg, *Travels*, I:195, 208; Sparrman, *Reize*, I:303–4, II:584; Swellengrebel Family Archives, Journal of Swellengrebel's Journey, 1776, p. 72; Borcherds, *Memoir*, p. 57; Barrow, *Travels*, II:401; W. Mackenzie, *Sketches of Travels in Southern Africa* (Edinburgh, 1824), p. 313; Pringle, *African Sketches*, pp. 177–78.

99. Sparrman, *Reize*, I:303–4; Pringle, *African Sketches*, pp. 177–78; R. Percival, *Cape of Good Hope*, p. 216.

100. C 206, Resolutions, 21 August 1792, p. 249. [Trans. note: This footnote is missing in the Afrikaans edition. I am deeply indebted to Mrs. Daleen van der Riet, Africana librarian at the J. S. Gericke Library, University of Stellenbosch, for locating a copy of *Trekboer* in their collection that contained the citation, written by hand, and initialed, by Dr. van der Merwe.]

101. Burchell, *Travels*, I:243–44; Barrow, *Travels*, I:105; Percival, *Cape of Good Hope*, p. 216.

102. BR 131, Scheeps and other Journals, H. Lichtenstein, 5 September 1805, p. 781; Lichtenstein, *Reisen*, I:119, 199.

103. 1/STB 20/1, Minute Letters, Stellenbosch Landdrost to Governor, September 1739; C 123, Resolutions, 23 March 1745, p. 121; F. Kersten, Memorandum on the Condition of the Colony in 1795, in Theal, *Records*, I:167–75; Testimony of C. H. Oliver, 8 December 1826, in Theal, *Records*, XXIX:478.

VI

Dispersal in the Interior and the "Decline" of Civilization

Fears for the Future

Sympathetic European visitors and concerned colonists during the eighteenth century frequently expressed the fear that as time went on the farmers in the interior, cut off from all civilizing influences, would degenerate and decline into barbarism. As early as 1713 the Drakenstein Church Council made known their concern for the colonists' future. They complained that they had already gone for six years without a minister "as a result of which this joyful community, now consisting of about seven hundred souls, large and small, becomes more morally degraded and destroyed as each day passes, and should it continue, they will finally decline to the level of the Hottentots."[1] Thirty years later a visiting commissioner, the well-known Baron van Imhoff, declared that on his journey through the interior he

> had perceived with astonishment and regret, how little public worship was carried on there, and also in what great unconcern and ignorance a large portion of the rural population lived their lives, knowing little or nothing of religion, so that they now more closely resemble an assemblage of uneducated pagans than a colony of Europeans and Christians.

The ministers he questioned regarding the cause for this state of affairs mainly attributed it to the great expanses of the outlying districts: the most distant colonists lived a three day's journey, or farther, from the church and this made regular attendance difficult.[2]

The establishment of a church in Tulbagh only temporarily

193

improved the situation. The farmers' migration into the interior continued unceasingly and very soon the vanguard were again beyond the church's reach. According to some, this rapid expansion of the colony was the cause of all evil. Thus Cloppenburg, for example, suggested in 1769 that measures be taken to halt the further expansion of the colony and maintain the colonists "as good citizens, since the far expanse was clearly a land of great degeneration in religious faith, in obedience, and consequently of good behavior among the current generation already, and future generations threaten to decline into total barbarism."[3]

Until the end of the eighteenth century the nearest church for farmers in the northern and northeastern border districts was at Tulbagh. Understandably, there were many who thus could not attend the church on a regular basis. Sometimes years went by, reported van Plettenberg in 1778, before the farmers of Camdeboo had their children baptized; "subsequently they must grow up as in the wild, without being able to obtain a solid foundation in religion. Nothing more may thus be expected from such an upbringing than all manner of wickedness."[4] In the same year the Camdeboo farmers lamented the decline of religion in the interior districts. They attributed this to the fact that they had to make do without teachers and ministers, "so that a large part of the young people grow up like dumb oxen, and we have had no opportunity to teach them the first principles. We shudder to think that when they have become older they will not be able to turn from their ways."[5]

About ten years or so later Mentzel, in his famous description of the Cape, expressed his opinion that as long as the Company continued to allow Europeans to settle in the colony who could mix with the colonists that were already here, and could civilized them, there was yet hope for the future.

> But after many years, should the land be populated only by native-born Africans, so that the Company would not find it acceptable to allow in more foreigners or Europeans, one must fear that the African nation would degenerate and become savages. Their nature is wild, their education bad, their thoughts base, and their conduct ill bred. One could presume, not without reason, that in time they might become more immoral, more

rude, and behave worse than the Scots, the Wends, and the Scythians of old.[6]

Around the same time Hendrik Swellengrebel, the son of the governor with the same name, observed that the farmers "lapse into slothfulness by living an idle, nomadic existence in the country, and with this lose all social virtues."[7] The same pessimism was also found among other native Afrikaners at the end of the eighteenth century, who saw in the poor economic conditions especially a reason for the social decline. "Agriculture," explained Cloete, the writer of a daily account of Swellengrebel's journey, "is a small diversionary activity that is more and more in decline among these fellows (the frontier farmers from Camdeboo). They keep their young men out on the hunt rather than in the fields and one can foresee a time when they will have become *an entirely wild and savage people.*"[8]

The drafters of the famous burgher petition of 1784 did not paint a rosy picture of the future. They perhaps deliberately exaggerated a little in order to improve the chances for the redress of their grievances. Limited marketing opportunities and an uncertain market for agriculture products, so it was claimed, stimulated the farmers to migrate into the interior, where they might earn a livelihood by livestock farming and hunting. These circumstances disturbed the petitioners and made them uneasy about their children's future,

> for while they face a total corruption of morals from life in such a wild country, and consider their beloved offspring, who they otherwise could extend a blessing under the kind hand of Providence, a totally rude nation, which quite possibly could become a dangerous region of the colony, such as the Bushmen and Hottentot lands are now. There one might legitimately ask the honest question "Can the nation by this time offer any subsistence to people among whom there is no civilization? What shall it do with those who mix with Hottentots and Xhosa?"[9]

The fear had been expressed a few years before (1779) that the colonists would mix with the natives, which could become dangerous for the colony, "for the young men, who have no means of establishing themselves, must remain unmarried, or spend their time among the Hottentots. From such mixing as time goes on an offspring can be born, who shall likely be feared worse than the Bushmen and Hottentots at present."[10]

The nervous anxiety over the future of the white race in South Africa, which repeatedly seized Cape colonists and European travelers in the eighteenth century, was evidently not altogether unfounded. But luckily their pessimistic expectations were not realized. In the face of this imminent ruin, which filled so many with such uneasy anxiety about the future, there was born in the eighteenth century a new nation, appearing like steel out of the crucible. But it is indeed a miracle that this pioneer population, living for two centuries in the most primitive conditions, having hardly any contact with the civilized world and constantly surrounded by barbarians, did not succumb to barbarism and mix with the natives. How then did this happen?

Education in the Country Areas

The influence of the social institutions that we look to today to preserve and disseminate our culture was very limited in the rural areas until well into the nineteenth century. The school, the most powerful social institution of this century, was more or less unknown. Indeed, even the establishment of schools in the outlying districts was nearly impossible because farms lay so far from each other and parents needed their children at home. It is true that in the nineteenth century small village schools developed here and there, but they were of little importance to the country people. Rural children generally received instruction from itinerant schoolmasters who were hired privately by the children's parents.[11]

These teachers, who occupied a relatively contemptible position in society, moved around from farm to farm throughout the colony. They usually never remained in the same place longer than six months or a year, and in this time they had to complete the education of their pupils. These so-called teachers were often discharged sailors or soldiers, who themselves could barely read and write and in many respects were a disgrace to their profession. The wages that they received were mostly paid *in kind*, and were so meager that there was no inducement for a better class of teacher to settle in the country areas.

Apart from the lack of suitable teachers the scarcity of textbooks and readers made education difficult in the old days. Beginners used an ABC book with illustrations of Adam and Eve. The follow-up to this was the "A.B.-jab," which contained two-syllable words. The well-known spelling book *Trap der Jeugd* [*Steps for Young People*] came after that. Generally children were taught only how to read, write, and count, as well as receiving a little religious instruction. They had to know the Lord's Prayer, the Ten Commandments, and the Twelve Articles of Faith. They also studied the catechism. More than this the parents regarded as unnecessary, although everyone insisted that their children receive at least this minimum instruction. Thus, although few rural children enjoyed an advanced education, there was very little total illiteracy.[12]

The value of this meager instruction, that was intended for little more than preparation for confirmation, must not be underestimated, however. It at least prepared the people to read their Bibles and thus perform the personal duties that their Protestant faith prescribed.[13]

Family Devotions

Although books were scarce in the countryside and the farmers did not read—a reproof that could still be made today—almost every family at least possessed a Bible. In 1820 Mary Moffat, wife of the famous missionary, wrote from Beaufort West to her parents, "I think I never saw so many fine-looking Bibles in my life as since I came to Africa. They [the Boers] seem to have a particular pride in them."[14] Besides the massive State Bible, that was looked after with the greatest care, there could also be found in most homes the hymns of Willem Sluiter, the Heidelberg Catechism and a condensed, simplified version of the Heidelberg Catechism known as the *Kortbegrip* "Synopsis" of Helmbroek. Some families also owned one or another book of sermons. The most well known of these was the *Donderslag der Goddelozen* [*Thunderbolt Against the Ungodly*] by Jodocus Lodenstein, which contained sermons with comments and prayers for every day of the year and all possible special occasions. In other books of sermons, such as *Berhardus Smijtegelt* for example, a biography of the author is also given.[15]

All the spiritual culture that could be found in the rural areas before the Great Trek emanated from the State Bible. This holy book was the torchbearer for Christian civilization in South Africa for two dark centuries, when the social institutions, which today lay claim to it, still had little influence. Yet today the head of a family will frequently ask one of the children to bring *the Book*. Often one also still hears talk of *boekevat* ["taking the book"], by which is meant family prayers or worship. Thus, Afrikaans language usage provides historical information about a long period in our people's history, when the Bible was the most important, often the only, book with which the farmers were acquainted and their library was limited to just what was necessary for the holding of family devotions.

The church certainly did its best to preserve the faith and to guard against the decline of civilization. Ministers in the interior districts sometimes displayed extraordinary zeal in going to visit their members, who lived in the farthest corners of their vast parish, and holding church services in outlying areas when they could bring together enough people.[16] The church's ministerings in the interior could not be very fruitful, however, because of the scarcity of churches and ministers, particularly during the eighteenth century. Undoubtedly the farmers' personal and daily use of the Bible contributed more to awaken and maintain their piety than the church did, which for many long years could do little for its dispersed flock in the interior.

Family worship occupied a very important place in the life of the pioneer family. Daily work did not begin before everyone had assembled for family worship. A meal could not begin or end without a prayer being offered at the table. This was the duty of one of the children, usually the youngest, but it was sometimes performed by the head of the family as well. If one of the grandparents was present, they were often asked to say grace or to give thanks. In the evenings after supper the family Bible appeared again on the table. The household remained seated, while slaves and

Khoikhoi, who also had to be present most of the time, squatted down alongside the wall. The father first read a piece from the Bible and perhaps something else from one or the other book of sermons and offered a prayer. After that everyone sang a hymn together, which they usually knew by heart. On Sunday mornings a special service was held.[17]

The Piety of the Farmers

The rural population of a century or two ago could certainly be described as a devout people, in spite of the scarcity of churches and ministers. Their piety, veneration for the minister and respect for the church were commented on by several travelers. Even on the colony's borders an itinerant missionary could always count on a good audience if he had the desire to hold a church service.[18] "It is to be admired," wrote W. von Meyer in 1841, "how much the African farmer is inclined toward piety, although he is isolated, has enjoyed little instruction and seldom receives spiritual ministering. There are certainly no people in the world who are so truly God-fearing as the Afrikaner."[19] Around the same time Chase stated, "With much truth we may describe the inhabitants of the Cape Colony at large as a serious and religious people, and especially with reference to that portion forming the most considerable part of the Community, the Dutch Boers, who are deeply imbued with strong sentiments of genuine piety, and are consistent members of the Christian Church."[20]

   In spite of their wide dispersal in the interior the farmers jealously held on to their church, which was slow to follow them on their long wanderings in the wilderness. The great distance that the old pioneers lived from "God's House," much greater than the distance from the market, made them unhappy in their isolation. Listen, for example, how the farmers of Camdeboo in 1778 complained about their fate in the language of the State Bible: ". . . how many dejected and shattered souls sigh, and long to approach the house of the Lord our God with lifted heads, and to take pleasure in God's law day and night, and to partake of the Lord's Supper, in the Christian manner, which opportunity is sometimes denied to him."[21]

Everyone wanted to have a church "where we can invoke the name of the Lord our God together," and an ordained minister in their neighborhood, "in order that God's acre in these widely spread-out districts is one day plowed." This great longing was repeatedly made known in numerous petitions to the governor and council as well as by verbal requests to visiting officials. That the farmers were serious in their requests appears evident from the willingness that was constantly displayed in the collection of individual contributions for the funds necessary to build the parsonage and church.[22] Up to the present day, the church remains the most substantial and prominent building in every small village, and a minister is treated with great respect by the rural population.

The few churches there were in the early days were attended regularly and dutifully by the farmers living in the vicinity, although sometimes they had to travel a few hours in a horse cart in order to reach one. Naturally those that lived farther away could not attend public worship on a regular basis, but they still visited the church now and then, because they were very insistent on the administrations of the sacraments. When de Mist and his company visited the Cold Bokkeveld in 1803, on the farm of the field-cornet Hugo, they began to doubt whether it would be possible to proceed any further on their journey over the mountains. The field-cornet's wife then tried to reassure them by recounting how every year a few weeks after her confinement she was in the habit of riding on horseback with her baby on her arm over the mountains to Roodezand's church in order to have the infant baptized.[23] The frontier farmers often had to wait until they could once again travel to the Holy Communion service before having their children baptized. Then it sometimes happened that the child was already a year old, or that more than one child was baptized at the same time.

Although the overwhelming majority of the farmers were unable to attend church regularly, they still did their best to be present one or more times a year for Holy Communion. Usually this also provided an opportunity for the confirmation of the young people and the admission of new members into the church. There were frequently a few hundred wagons gathered together round the church at these communion services. Some farmers lived more than two hundred miles from the church, and the journey lasted from fourteen days to three weeks, but they attended Holy Communion all the same unless prevented from doing so by drought, locusts, or commando service.[24]

The Influence of the State Bible

Naturally there would also be those who remained indifferent toward religion. Undoubtedly mealtime prayers, family worship, and church attendance became a mere social custom in some cases. But this still should not detract from the enormous social significance of the piety of the farmers in general. In pioneer society the church steeple was the inexorable center of social life. Each child had to be baptized and everyone had to become a member of the church. Parents would not give their consent to a marriage before a son or a daughter was confirmed. The moral standards accepted by the society as a whole also found their theoretical basis in the State Bible. In many cases, obviously, there was some deviation from this, but they would not receive social sanction. The absolute validity of Christian ethics was never seriously questioned in the rural areas, and little intellectual influence to undermine their validity reached the farmers in their isolation.

We possess almost no statistics that can throw light on the ethical character of the rural population, but travelers generally received a very favorable impression of their morality.[25] Although up to the end of the eighteenth century the frontier farmers had to travel to Cape Town for the performance of marriage ceremonies—which was a perpetual problem—they remained very much committed to having their marriages legally sanctioned. Girls usually married young and a virtuous married life was the custom. Serious social evils, such as drunkenness, were nearly unknown in the country. A considerable percentage of the farmers were teetotalers and those who drank usually did so in moderation. There was no lack of social order in spite of the fact that the arm of the law did not reach very far from Cape Town. Stories that show an extreme lack of respect for life and property—such as those told in the pioneer history of some other young nations—do not blemish the history of the Cape Colony. In the early days, we are further assured, it was not necessary to draw up a written contract of purchase or to sign a receipt. A man's word was his honor. Travelers were quickly struck by the exemplary family life of the old pioneers as well. Mention is repeatedly made of the respect that children, mindful of the fifth Commandment, displayed toward their parents.

For the most part the influence of the State Bible and the Church is to thank for the fact that the dispersal and isolation of the farmers in the wilderness did not result in the decline of public morals. On their migration into the interior the farmers were compelled to leave behind much that belonged in the home of a civilized person, but yet travelers were often surprised to encounter so much culture on the frontier, in spite of the limited possibilities of acquiring it. Perhaps farmers during the eighteenth and nineteenth centuries even gained more of a spiritual culture than they lost.[26]

Miscegenation with the Indigenous Peoples

The fear expressed now and then that the colonists in the interior would interbreed with the indigenous peoples was also exaggerated.

That miscegenation occurred between whites and the indigenous peoples at the Cape is evidenced by the development of the Cape Coloured in the Western Province and the Griquas and other mixed-race groups in the northern border districts. Unfortunately we possess no statistics to indicate how much of that miscegenation was the result of legal marriages with female slaves, Khoikhoi and Coloureds, and how much extramarital sexual relations between whites and Coloureds contributed to it. It is also impossible to solve the delicate problem as to the extent that the sailors and soldiers in the seaport, Company servants who deserted inland, and the colonists themselves were variously responsible.[27] There is no doubt whatever that no causal connection exists between the miscegenation with the indigenous peoples and the farmers' dispersal into the interior. As proof it can be argued—as the Council of Policy rightly explained in 1786 in answer to the petition of 1784—that the miscegenation with the indigenous peoples had already begun long before the commencement of the expansion into the interior.[28]

Indeed, the risk of whites interbreeding with the indigenous peoples even diminished with the passage of time.

During the first years of the settlement, when white women

were still very scarce, there was evidently a great deal of sexual intercourse with the slave women.[29] Regular marriages between whites and baptized female slaves and Khoikhoi were even concluded, and they had full social approval and religious sanction. The mixed-race offspring from such marriages were accepted into the white community without hesitation.[30]

Fortunately for the purity of the blood of the white race in South Africa, colonists born in this country radically altered their liberal attitude toward the non-European population groups relatively quickly. As time went on race and skin color became more important in determining social relationships, and gradually there developed a strongly expressed color prejudice, which formed an unbridgeable social gulf between white and nonwhite. Initially perhaps this color prejudice was based on aesthetic grounds, but undoubtedly it was also connected with the knowledge that the whites would not be capable of holding their own against the indigenous peoples if they treated them as equals.

Prejudice against the mixed-race and indigenous peoples did not yet exist at the beginning of the eighteenth century. It developed by degrees in the course of time and by 1770, when the colonists came into contact with the Bantu, it was strong enough to prevent miscegenation with that race. Before the end of the eighteenth century social equality between whites and people of color was regarded as something evil. A dark skin and woolly hair were enough to create an unbridgeable social gulf even though no other differences existed.

In the documents from the pens of colonists of the last half of the eighteenth century *Christian* and *Heathen* were synonymous at that time for "white" and "black." The indigenous peoples were the descendants of the cursed children of Ham and accordingly it was their lot and destiny to remain socially inferior. To place a Christian on an equal basis with a heathen was thus contrary to the will of God.[31] Moodie states that the question of what was to become of baptized slaves and Khoikhoi was a problem that gave cause for concern to many, although the Bible reference to the many mansions in the Kingdom of Heaven brought a little relief in that it established the possibility of eventual segregation in the afterlife![32]

Thus one can perhaps conclude that the attitude of the Cape

colonists toward the indigenous peoples was grounded in the Old Testament, which would also explain why this peculiar color prejudice developed as the Old Testament took a greater hold on the farmers. On the other hand it is just as possible that the pioneers defended their attitude toward the indigenous peoples by appealing to the authority of the Bible, when in truth it was based on other grounds.

NOTES

1. Church Council of Drakenstein to Classis, Amsterdam, 26 March 1713, in C. Spoelstra, *Bouwstoffen voor de geschiedenis der Nederduitsch-Gereformeerde Kerken in Zuid-Afrika* (Amsterdam, 1906), I:128.

2. C 121, Resolutions, 19 February 1743, pp. 135–36.

3. AR, Accession 242, J. W. Cloppenburg, Notes and Remarks, p. 1.

4. C 156, Resolutions, 1 December 1778, p. 391.

5. C 1269, Petitions and Requests, Petition from Camdeboo, 24 March 1778, p. 429.

6. F. Mentzel, *Vollständige und Zuverläszige . . . Beschreibung . . . des Guten Hoffnung* (Glogau, 1785, 1787), II:185.

7. Swellengrebel Family Archives, H. Swellengrebel, Notes (undated).

8. Swellengrebel Family Archives, Cloete Journal, 1776, p. 27; cf. Commission for Stock Raising and Agriculture, 20 November 1805, in G. M. Theal, *Belangrijke Historische Dokumenten over Zuid-Afrika* (Cape Town, 1896, 1911), III:428.

9. C 1283, Petitions, Reports, etc., Burgher Petition of 17 February 1784.

10. AR, Accession 40/548(a) (Nederburg: Petition, 10 July 1779).

11. CO 49, Court of Justice, Report of the Circuit Court, 1813; CO 60, Court of Justice, Report of the Circuit Court, 1814; GH 23/8, Dispatch Book, R. Bourke to Huskisson, 19 May 1828, new pp. 374–75; Rev. G. Thom to Lord Charles Somerset, 26 January 1821, in G. M. Theal, *Records of the Cape Colony, 1793–1831* (Cape Town, 1897–1905), XIII:386–89; W. J. Burchell, *Travels in the Interior of Southern Africa* (London, 1822), I:199; C. Latrobe, *Journal of a Visit to South Africa in 1815 and 1816 with some account of the Missionary Settlements of the United Brethren near the Cape of Good Hope* (London, 1812), p. 252; H. Lichtenstein, *Reisen im südlichen Africa in den Jahren 1803, 1804, 1805 und 1806* (Berlin, 1811–1812), II:127.

12. CO 41, Court of Justice, General Report of the Commission of Circuit, 1812; CO 49, Court of Justice, General Report of the Commission of Circuit, 1813; CO 60, Court of Justice, General Report of the Commission of Circuit, 1814; CO 2744, Letters Received from Worcester, Truter to Colonial Secretary, 5 November 1833; Latrobe, *Journal*, p. 252; C. J. F. Bunbury, *Journal of a Residence at the Cape of Good Hope, with Excursions into the Interior* (London, 1820), p. 183; C. De Jong, *Reizen naar de Kaap de Goede Hoop* (Harlem, 1802–1803), p. 133–34; "Remembrances of L. C. de Klerk," in Preller, *Voortrekkermense*, I:256–57; E. G. Malherbe, *Onderwys en die Armblanke* (Stellenbosch, 1932), p. 18.

13. Malherbe, *Onderwys en die Armblanke*, p. 18.

14. J. S. Moffat, *The Lives of Robert and Mary Moffat* (London, 1890), p. 74.

15. AR, Accession 1914, (van Plettenberg Collection, no. 28), O. G. de Wet, Daily account of a Journey (1778), p. 33; Commission for Stock Raising and Agriculture, 20 November 1805, in Theal, *Belangrijke Historische Dokumenten*, III:428; P. B. Borcherds, *An Auto-Biographical Memoir* (Cape Town, 1861), p. 55; J. Barrow, *Travels into the Interior of Southern Africa* (London, 1823), I:82; Bunbury, *Journal*, p. 183; "Remembrances of L. C. de Klerk," in Preller, *Voortrekkermense*, I:256.

16. Burchell, *Travels*, II:154; J. C. Chase, *The Cape of Good Hope and the Eastern Province of Algoa Bay* (London, 1843), p. 136; CO 60, Court of Justice: General Report of Circuit Court, 1814.

17. Commission for Stock Raising and Agriculture, 20 November 1805, in Theal, *Belangrijke Historische Dokumenten*, III:427–28; E. C. Godée Molsbergen, *Reizen in Zuid-Afrika* (The Hague, 1932), II:266; Borcherds, *Memoir*, p. 55; W. von Meyer, *Reisen in Süd-Afrika wärend der Jahre 1840 und 1841* (Hamburg, 1843), p. 171; Barrow, *Travels*, I: 82; Chase, *Cape of Good Hope*, p. 135; C.R. Baynes, *Notes and Reflections during a Ramble in the East, etc.* (London, 1843), p. 35; T. Smith, *South Africa Delineated* (London, 1850), p. 17.

18. CO 41, Court of Justice, General Report of the Circuit Court, 1812; Bunbury, *Journal*, p. 183; Borcherds, *Memoir*, p. 55; S. Broadbent, *A Narrative of the First Introduction of Christianity Amongst the Barolong Tribe of Bechuanas, South Africa: With a Brief Summary of the Subsequent History of the Wesleyan Mission to the Same People* (London, 1865), p. 136.

19. Von Meyer, *Reisen*, p. 171.

20. Chase, *Cape of Good Hope*, p. 135.

21. C 1269, Petitions, Reports, etc., Petition from Camdeboo, 23 March 1778, p. 429.

22. C 1428 and 1429, Letters Disp., Willem Adriaan van der Stel to Patria, 1 April 1703; 1/SWM 1/1, Minutes, 7 May 1751, p. 215; AR,

Accession 1914 (van Plettenberg Collection, no. 28), O. G. de Wet, Daily account of a Journey (1778), p. 33; C 1813 and 1814, Letters Disp., Cape Government to Patria, 19 April 1786; Commission for Stock Raising and Agriculture, 20 November 1805, in Theal, *Belangrijke Historische Dokumenten*, III:428; CO 2568, D. J. van Ryneveld to Tulbagh Landdrost, 20 November 1808; CO 41, Court of Justice, General Report of the Commission of Circuit, 1811–1812; CO 336, Commissioner-General, A. Stockenstrom to Plasket, 1 June 1825, new pp. 112–15; 1/GR 16/52, Letters Disp., Civil Commissioner to Colonial Secretary, 26 March 1834; Chase, *Cape of Good Hope*, p. 135; A. Steedman, *Wanderings and Adventures in the Interior of Southern Africa* (London, 1835), I:168; C 1269, Petitions, Reports, etc., J. Kruger et al. from Camdeboo, 24 March 1778.

23. A. U. de Mist, "Dagverhaal," p. 101; Godée Molsbergen, *Reizen*, II:188; Lichtenstein, *Reisen*, I:216.

24. Journal of van Plettenberg's Journey (1778), in Theal, *Belangrijke Historische Dokumenten*, I:12; Barrow, *Travels*, I:83; CO 41, General Report of the Commission of Circuit, 1812; "Introductory Remarks," in R. Godlonton, *A Narrative of the Irruption of the Kafir Hordes into the Eastern Province of the Cape of Good Hope 1834–1835* (Grahamstown, 1836), p. 17 [Trans. note: vdM incorrectly gives the author of this book as C. A. Fairbridge and also gives an inaccurate title. See the Trans. Note in the bibliography for this book under Godlonton.]; Burchell, *Travels*, II:142; Chase, *Cape of Good Hope*, p. 135; Baynes, *Notes and Reflections*, p. 39; Smith, *South Africa Delineated*, p. 117; Steedman, *Wanderings*, I:167.

25. Cf. Lichtenstein, *Reisen*, I:268.

26. PSB, Mss. Germ. Fol. 879, J. A. de Mist, *Reisverhaal*; Commission for Stock Raising and Agriculture, 20 November 1805, in Theal, *Belangrijke Historische Dokumenten*, III:427–28; CO 49, General Report of the Circuit Court, 1813; 1/STB 20/2, Letters Rec., Landdrost Faber to Governor Tulbagh, 13 February 1770; Chase, *Cape of Good Hope*, p. 135; Bunbury, *Journal*, p. 183; Baynes, *Notes and Reflections*, p. 183; Lichtenstein, *Reisen*, I:120, 149, 173.

27. See K. P. Thunberg, *Travels in Europe, Africa and Asia performed within the years 1770 and 1779* (London, 1793), I:179; Sparrman, *Reize*, I:321; Mentzel, *Beschreibung*, II:183.

28. C 171, Resolutions, 19 April 1786, p. 424.

29. H. T. Colenbrander, *De Afkomst der Boeren* ([The Hague], 1902; reprinted, Cape Town: Struik, 1964), p. 119.

30. I. D. MacCrone, *Race Attitudes in South Africa: Historical, Experimental and Psychological Studies* (London, 1937), p. 42ff.

31. Janssens to de Mist, 12 May 1803, in Theal, *Belangrijke Historische Dokumenten*, III:209; J. A. de Mist, *The Memorandum of Commissary J. A.*

de Mist (Cape Town: Van Riebeeck Society, 1920),series I, vol. 3, pp. 256–57; Baron A. van Pallandt, *General Remarks on the Cape of Good Hope* (1803), p. 14; "Remembrances of Anna Steenkamp," in Preller, *Voortrekkermense*, II:30; Colonial Library, The Hague, Paravicini di Capelli, *Reizen.* [V.R.S., entry for 29 April 1803, series I, vol. 46, p. 49.]

32. J. W. D. Moodie, *Ten Years in Southern Africa* (London, 1835), I:158–59.

VII

The First Conflict between
Whites and Bantu

The initial contacts between whites and Bantu occurred at about the same time as the first expeditions, undertaken generally by private parties, that journeyed inland from the beginning of the eighteenth century. These expeditions took place with an eye both on the livestock trade with the indigenous peoples and elephant hunting. The livestock trade was forbidden "under penalty of life and limb" soon thereafter because the government held fast to its belief that the colonists, "the poor, and with respect to the Europeans, defenseless people," had been robbed of their livestock by force.[1] Then it was chiefly elephant hunters that penetrated into Xhosaland.[2] Nevertheless, the unlawful livestock trade was still carried on, hand in hand with the hunting. Such expeditions were undertaken so often that in 1752 Ensign Beutler from Swellendam could follow the route of the traders and hunters into Xhosaland.[3]

The Bantu were primarily hunters and cattle farmers. "Their greatest asset," reports the account of van Plettenberg's journey, "consists of cattle which they possess in the thousands," but they also practiced agriculture, although in a primitive manner. The ground was dug up with short sticks, and then pumpkins, watermelons, peas, beans, marijuana, tobacco, and grain sorghum were sown.[4] They thus demonstrated a greater interest in the land than the Bushmen and the Khoikhoi. In addition they stood much higher culturally and they were economically and politically much stronger than the other indigenous peoples. This explains why the Xhosa could offer such resolute opposition to white expansion into the interior.

While the stream of emigration from the colony flowed eastward, the eastern Bantu groups were moving westward. The Bantu migrated more slowly than the whites, probably because they had better land. Also perhaps because, in spite of greater numbers, they spread out in groups. According to the testimony of an old farmer whom Collins mentions, the Keiskamma River was regarded in 1738 as the western boundary of Xhosaland.[5] In 1752 Ensign Beutler corroborated this assertion, but according to his travelogue it appears rather that the foremost Xhosa chiefs lived a long way to the east of the Keiskamma River.[6] In 1756 the Fish River was designated as the Xhosaland border,[7] and during the 1770s the two streams of emigration came together in the vicinity of the Fish River. The foremost Bantu groups evidently crossed over the Fish River shortly after the first colonists had settled beyond Bruintjieshoogte. Be that as it may, by the end of 1778 van Plettenberg had met with various Xhosa chiefs west of the Fish River and a conflict between white and Bantu threatened even then.

Conflicts of Interest over Grazing Land

The colonists as well as the Xhosa were stock farmers with extensive herds. Very shortly after the first contact between the two groups there were clashing interests over grazing land. Van Plettenberg made the following observation in the daily account of his journey:

> The Xhosa, who in former times always remained about a day's ride to the east of the Great Fish River, have for a few years now advanced more and more toward us and at last have encamped with their herds of cattle on this side of the same river. The colonists regard this not only as a great nuisance and loss of grazing land but, because of the superior numbers or treachery of the Xhosa that they are beginning to fear, they were also forced to abandon the farms that they possessed on loan in the vicinity of the river.[8]

Up to then the government had always had difficulty in holding back the colonists, who constantly wanted to migrate further eastwards. Now, however, the border problem had changed. The Xhosa had penetrated into the colony with their livestock, and their presence made the isolated frontier farmers feel insecure. The governor recognized the necessity of subjecting *them* to a fixed frontier border. And it is obvious from his discussions with the Xhosa chiefs that van Plettenberg had had an eye on the conflicts that could originate over grazing land when he fixed his border in 1778. He came to an agreement with the chiefs to regard the Fish River as the boundary between the colonists and the Xhosa, and laid special emphasis on the provision that they must not come across the river with their livestock, nor make use of the open country to the west to graze them. This point is so important that it is better perhaps to let the documents speak for themselves in this connection.

The first two chiefs van Plettenberg dealt with were Koba and Ghodisa. The governor

> remonstrated with them over the illegality of their position on this side of the river, and he brought to their attention that if their people continued to make use of the region this would result in a severing of the friendship between them and the Honorable Company, that up till then had still come to their aid. He then issued a friendly warning. If they were inclined for this to happen, then for one thing they, together with their people and all their cattle, would have to abandon both sides of the Great Fish River, in order to allow the same river to be extended as a fixed line of demarcation between them and the possessions of the Honorable Company. Thereupon the chiefs countered that they would lose the harvest of corn they had sown and that was already sprouting if they obeyed the order immediately.

They promised, however, that as soon as the crops were gathered they, "with their people and all their cattle, will move away and abandon for all time the region on this side of the said river, and this promise they solemnized by shaking our hands and striking their chests."[9]

In his discussions with Jamosie, whom he had met a day's ride to the west of the Fish River with a following of men, women, and children, the governor stressed the same issues. This chief, who, "with a herd of cattle had settled down on this side of the Bushmen River mountains, so that we suspected that he and his people would likewise seek to intrude more and more among our people," was also notified by the governor "that he and his followers would have to remain on the other side of the Bushmen River mountains and thus outside of our colonial districts, and not use the grazing lands within these districts. This too he promises to honor."[10] Shortly after returning from his journey, van Plettenberg wrote to his friend Swellengrebel of his negotiations with the Xhosa chiefs "who encamped with a large number of cattle on this side of the river" and he reported that they had promised "they will cross over the Great Fish River, and remain there, without coming to graze their cattle on this side."[11]

From the foregoing description it will appear very obvious that the Xhosa presence with their livestock to the west of the Fish River, which placed the farmers in the pasturage at a disadvantage, was the essential issue around which the governor's border demarcations with the Xhosa revolved. It is thus somewhat peculiar that Dr. Theal failed altogether to mention this point in his standard work on the history of South Africa. He gives a summary of the story of van Plettenberg's journey, which we cited above, but from his account in regard to the border demarcations with the Xhosa it is not clear why the governor wanted to keep the Xhosa east of the Fish River:

> Koba and another petty Captain named Godisa promised the Governor to respect the Great Fish river as a boundary between themselves and the white people, and to recall their followers who were on the western side as soon as the crops which those people had planted were gathered. . . . At the Bushman's river a petty Xosa captain named Jamosi was met. He promised to return at once to the Country from which he had come, and was thereupon gratified with a trifling gift.[12]

On the other hand Dr. Theal furnishes information on the relationship between white and Bantu at this time for which I could find no documentary support in spite of the most careful research in all possible sources. "The farmers about the Sneeuberg were found to be suffering severely from the depredations of Bushmen. Their chief request was to be provided with a landdrost and a clergyman . . . ," relates Dr. Theal. Later he speaks of the farmers living along the Great Fish River: "They made the same request as those of the Sneeuberg, and their grievances were identical, if Xosas be substituted for Bushmen." On the same page he writes further, "The Xosas, who were addicted to lifting the stock of the European graziers, had only recently moved so far westward."[13]

In van Plettenberg's account there is no indication that farmers had complained to the governor about Xhosa robberies. If the Xhosa had truly stolen the farmers' livestock in 1778, the governor would surely have heard about it. One can also expect that the travel journal would have made mention of these offenses at the time and that the governor in his discussions with the Xhosa chiefs would have raised the issue. This appears not to have been the case. Only the question of grazing land was discussed. In his letter to Swellengrebel, van Plettenberg also makes no mention of Xhosa stock thefts.

On a later occasion the governor again referred to his frontier journey. He spoke with admiration about the simple lifestyle of the frontier farmers and explained that they would have been extremely happy "if their welfare was not continually disturbed by the marauding of the so-called Bushmen Hottentots and the injury that was caused for some time here and there by the Xhosa."[14] From this it appears that the governor did not place the Bushmen and Xhosa as neighbors on the same level.

What the governor meant precisely by *injury*, is not altogether clear. But if the Xhosa had truly stolen the farmers' livestock, there is no reason why he would have described their thievery with the word *injury*. Presumably he was primarily thinking of the great herds of cattle that the Xhosa, who traveled in groups, grazed even on the colonists' farms.[15] It is nevertheless also possible that the frontier farmers could have found troublesome the Xhosa visits, which they made in order to be entertained and to receive presents. Lichtenstein tells us that the Xhosa, who held the view that if you lived in peace with your neighbors you also had the right to visit them and to inquire how they were, sometimes hung about the farms in large numbers for days and even weeks, while the farmers had to entertain them to preserve the peace.[16]

Since support for Dr. Theal's standpoint is nowhere to be found in the sources of the time, and inasmuch as he does not give any documentation of his assertions in the customary manner through the use of footnotes, I cannot for the moment reconcile myself with his representation of the matter with regard to this point. If, with the greatest modesty, I might differ with such a renowned authority on our history, then I would summarize the situation on the eastern frontier at the end of 1778 as follows: after the frontier farmers and the Xhosa had already lived in close contact with one another for a considerable time to the west of the Fish River, the governor decided that the whites should remain in possession of the fertile ground to the west of the Fish River, which they had evidently occupied before the Xhosa. In order to avoid disputes provoked by questions over grazing land in the future, the Xhosa had to vacate the land to the west of the Fish River. There were then two conditions present for a future conflict. The Xhosa had destroyed the farmers' grazing lands with their large herds of cattle. The isolated frontier farmers felt insecure surrounded by the great number of Xhosa, which they also now began to fear and distrust. For this reason some farmers abandoned all the farms they owned along the Fish River. If the Xhosa did not follow through with their promises to leave the land west of the Fish River voluntarily, there could have developed among the farmers a desire to compel them to do so.

The First Open Hostilities

In spite of the promises made to the governor, and under all sorts of pretenses, the Xhosa refused to move back across the Fish River. In April 1779 they were still on the site where the governor had visited them. They even spoke of beginning to sow corn once more. The farmers knew what this would lead to and petitioned Adriaan van Jaarsveld, the field-corporal of Camdeboo and Agter-Bruintjieshoogte, to drive the Xhosa out of the colony. He called in the Stellenbosch landdrost for help, but did not get any assistance from the government.[17] It appears, however, as if no open hostilities between white and Bantu occurred in the Stellenbosch district up to the beginning of December 1779. The field-cornets' reports that remain preserved are silent regarding the Xhosa, and only the disturbances with the Bushmen are referred to in the discussions over the incoming field-cornets' reports in the meeting of heemraden and military officials held on 7 December 1779.[18]

It was less peaceful, however, in the Swellendam district. According to a declaration from Field-Corporal Stephanus Scheepers, in June 1779 the Xhosa murdered the herdsman on his livestock farm along the Bushmans River and took away sixty cattle.[19] On 19 August they killed H. C. Janse van Rensburg's cattle herder in his absence and drove away half of his herd, a few of which later came back on their own. On another farm in the neighborhood a couple of Khoikhoi were murdered by the Xhosa.[20]

On 25 August 1779 the Swellendam landdrost wrote to the governor,

> whereas a succession of complaints have been brought in by the inhabitants on this side of the Bushmans River, with regard to the disturbances with the Xhosa, I have, under Your Excellency's favorable approbation, appointed a field-corporal to serve there, in order to check and prevent as far as possible, all disorders and rash undertakings by both sides.[21]

The precise nature of the "disturbances" was not stated. The landdrost would verbally relate more to the governor in September on this matter and then ask for instructions. From the Swellendam logbook as well there is no indication of what led to Lucas Meyer's appointment. It only records that a decision was taken to charge him "with seeking an amiable way in which to oust the Xhosa settled on this side of the Bushmans River."[22]

Until 25 October, according to the logbook, nothing further of importance occurred in this matter. At the heemraad meeting held on this date, however, the landdrost put the case that "very many complaints were being received daily about the conduct and the tyranny of the neighboring Xhosa, although without anyone being able to rely much on any of them since nearly all the messages varied greatly one from the other." He subsequently asked the members of the gathering whether any of them could say anything with certainty, so that the government could be made aware of it. Since no one, however, "knew anything different except further rumors," no decision was then taken and the members were asked to let the landdrost know the moment they heard anything of importance.[23]

Three days later the landdrost received a written communica-

tion from Field-Cornet J. H. Potgieter that some farmers on the
Bushmans River had fled from the Xhosa. The landdrost found
that no "well-grounded reasons for the cause of this" were stated
in the letter, and he forwarded a copy of it to the governor.[24] This
extremely interesting note was written on 27 September. The field
cornet relates that eighteen farmers, whom he lists by name, had
fled from the Xhosa. Furthermore, he wanted to make known,

> that I shall keep the men at my home. My Lord I cannot know
> what more that nation is up to in my district. My Lord I should
> have already done a report but I wanted first to speak with the
> men in order to report to you My Lord on the matter. My Lord
> I should rightly be coming myself to give My Lord the report
> but I have worn out my horses in riding to learn the truth. My
> Lord this nation is only waiting until the men are first at drill
> then they would have it their own way so that I do not know
> My Lord what they have in mind. The men who have fled have
> also already had their livestock stolen. . . .[25]

From this it appears that the field-cornet, who rode his horses
to exhaustion in order to discover the cause of the farmers' flight,
did not want to risk stopping them from fleeing because he dis-
trusted the Xhosa. It also appears as if the farmers were afraid the
Xhosa would attack them by surprise and believed they were only
waiting for a favorable opportunity to do so. From the letter one
could not infer that the Xhosa were guilty of committing any serious acts of violence. The vague references to stock theft at the end
of the letter make one suspect that it could not have assumed great
proportions, and in any case was not the principal cause for the
farmers' flight. For one reason or other, however, a strained state
of affairs had developed. It was exacerbated through fear and distrust and gradually became so serious that charges arose that the
Xhosa were "even already" beginning to steal livestock.

A declaration that Jan Scholts, field-corporal at the Gamtoos
River, laid before the Swellendam secretary on 5 November 1779,
throws light on the further development of these disturbances on
the eastern frontier. He had asked permission from the landdrost
to make a journey into the interior, and this was granted. At the
same time he was directed to gain information with regard to the
disputes between the colonists and the Xhosa. On 10 October he

came upon the farmers who had fled from the Xhosa on the Swartkops River and they told him they "were fleeing from beyond the Swartkops River, as they testified to Potgieter, since the Xhosa had driven them away, burned their houses, and taken away many of their cattle."[26]

From the above statements it appears as if between June and October 1779 Xhosa stole livestock, at least on a small scale, from the farmers living along the Bushmans River and were also guilty of committing other acts of violence. Furthermore, it is interesting that the Xhosa alleged that they were motivated to commit those acts by the fact that the Gonaqua Khoikhoi stole their livestock. The Gonaqua Khoikhoi lived among the farmers and the Xhosa suspected that they acted as the farmers' henchmen.[27]

The first open hostilities between white and Bantu took place toward the end of 1779. They occurred in the Stellenbosch district, where apparently until then the farmers had not suffered any Xhosa violence. On 13 March 1780 the Stellenbosch landdrost, De Wet, notified the governor of the new hostilities. He wrote that from two letters he had received two days before from Josua Joubert, field-cornet from Camdebos River to Bruintjieshoogte,[28] he learned for the first time

> how for the last three months some open hostilities had broken out between the Xhosa beyond Bruintjieshoogte and the colonists living there. The Xhosa had stolen a substantial number of cattle from the inhabitants and in the reciprocal attack some of the inhabitants were wounded by the force collected by the Xhosa, who were themselves then put to flight, leaving behind eight or nine dead. In order to go against the continued threat of violence from the Xhosa, a strong commando was ready to depart. The outcome of this action, so it appears to me, will determine whether the Xhosa are to be dislodged with force, or whether it will be necessary for the inhabitants themselves to clear out of the country.[29]

The "strong commando" that De Wet mentioned must have gone out early in 1780. At about the same time a commando also marched out from Swellendam against the Xhosa. According to the declaration of Burgher-Adjutant Pieter Hendrik Ferairre it took place under the following circumstances.[30] The Langkloof burghers, some farmers living further inland, some burghers who came under the jurisdiction of Stellenbosch, and Commandant Hendrik Potgieter, all insisted that Ferairre raise a commando in order to take back the livestock stolen by the Xhosa. Ferairre then sent a letter to the Swellendam magistrate describing the situation, but the letter was delayed along the way and was late in arriving. This was why the Swellendam commando marched out against the Xhosa without instructions, permission, or even notification.

In the field Ferairre's commando met up with the Stellenbosch commando, after the latter had already engaged in fighting with the Xhosa. Ferairre was then compelled to take command of the combined commandos. The two commandos united west of the Fish River but then penetrated deep into Xhosaland. Since Ferairre had called up his burghers simply and solely to recover stolen cattle, he ordered them not to shoot before he gave the command. This he would only do if forced by the most urgent necessity to do so. He had, in point of fact, not given the command to fire when the commando encountered the Xhosa. But "when it was clear that some arrows from the aforesaid Xhosa were coming down on them" the burghers began to shoot on their own. Many Xhosa were shot dead and a large number of cattle were seized. Among these there were animals that the Xhosa had taken from the farmers in the past.[31] On the suggestion of the Stellenbosch commandant—Cornelis Bothma, Hendrik Meintjies, Josua Joubert, and one Krugel—the livestock were then divided equally as booty among the commando members.[32] Ferairre spoke out against this, declaring

> that he had not gone to take cattle from the Xhosa but solely to see that the stolen ones be recaptured. Therefore he would absolutely not give his assent to any division, except only in so far as that while the herd was too large to remain together, everyone should take a portion of the herd under his supervision until such time as the government had disposed of the animals. He then gave orders to the men under his command to comply with his wishes.[33]

The Causes of the First Open Hostilities

When Landdrost De Wet sent news of the encounter between the frontier farmers and the Xhosa to the governor, he also enclosed the two letters of Josua Joubert from which he had obtained his information. To this he remarked,

> From the first letter it will be evident to Your Excellency, that these hostilities chiefly arose as a result of the violence and annoyances perpetrated by the inhabitants against the Xhosa. They complained to the field-cornet that Willem Prinsloo, Sr., had made himself master of some cattle that belonged to them, and also that Marthinus Prinsloo himself, or some of his companions, on their journey into Xhosaland, had murdered one of Chief Gaggabe's subordinates. Yet when confronted with these accusations, they delivered contrary accounts or their best excuses.

The landdrost was inclined to believe the Xhosa's story:

> It is not altogether improbable that these charges and accusations are more or less well grounded. On the one hand the Xhosa nature is vindictive. On the other hand, they are not barbarous enough to attempt such brutish and daring undertakings without just cause. The council is well aware, however, that all efforts with respect to a close examination of this case in order to uncover the truth, would be fruitless because of the extreme remoteness of the region. [34]

These accusations against the Prinsloos also came to the attention of Dr. Theal, but he was evidently inclined to doubt their validity: "It is impossible now to say whether these reports were true or not. They were put in circulation by men who were certainly biased in favour of the Kaffirs, and the explanations of the Prinsloos, if any were given, are not to be found." [35] This argument is not convincing, because it is highly unlikely that Field-Cornet Joubert, the man who led the first commando against the Xhosa, would be biased on the side of the Xhosa. What precisely De Wet's view was is difficult now to determine, but he was undoubtedly well-informed about the affair, because in 1778 he toured the eastern border together with the governor. Moreover, Joubert's contention that one or two irresponsible frontier farmers could cause the Xhosa to revolt is well founded when tested against the statements of persons living at that time and possessing intimate knowledge of frontier conditions.

In 1780 van Jaarsveld asked to be allowed to make peace with the Xhosa. He would return to them the livestock taken during the war, "because it is certain that the Xhosa are also people who have been robbed and brought to poverty, who also know well the feeling of never having cattle of their own. In such a way we could thus lessen the suffering under the occasioned oppression." On the other hand he supposed the Xhosa were also not exempt from all blame for the conflicts. It is, nevertheless, interesting to see how he arrived at the opinion

> that I have found many improprieties committed by the Xhosa. First, they have placed their homes on this side of the Great Fish River and there they have a region of land, which I now swear to have seen, which for them is abundant and much better than the Fish River. Second, they have set themselves apart in order no longer to submit to their ruler, and while even now their ruler seeks to make peace with us by returning our cattle, his people still appear insolent and have taken even more. Third, they profess that they have suffered many tribulations because of us but, nevertheless, shall not be delivered from such worries by once again returning to their own land, which is not very cooperative of them.

Van Jaarsveld further explains that he had carefully attempted to inquire into whether the allegations that the "harm" done the Xhosa were true, "but found that everyone on our part as well as on theirs seeks to avoid getting involved, and have not found one allegation that held up, excepting only that unconscionable deed that occurred on the commando of Barend Bouwer, and thus, what I have reported to the Landdrost holds good."[36]

In 1784 van Jaarsveld again requested that certain persons from Agter-Bruintjieshoogte be removed, otherwise the law-abiding inhabitants among them would suffer. "If the Kaffirs are stirred up as before, and similarly if the avaricious persons we mentioned above threaten them daily, so that we shall have to defeat the Xhosa again to have peace," he said, the consequences would be terrible for the colony.[37]

Among unsorted documents in the Cape Archives I found in 1932 yet another interesting report that sheds light on the causes of this first conflict with the Bantu. The document in question is not signed, but the handwriting, style, and contents create a strong suspicion that Adriaan van Jaarsveld is the author. The date is missing as well, but the report must have been drawn up after 6 August 1782—the earliest date mentioned in the document. It could not have happened very long after that date, however, because the document falls clearly within the time of van Jaarsveld's dispute with Field-Corporal Bothma,[38] against whom the following charges were brought:

> during the first unauthorized expedition into Xhosaland, made by the Prinsloos in the year 1776 (?), the first Xhosa revolt erupted when the colonists brought back a large herd of cattle. This was followed shortly thereafter by a rumor from the Xhosa, and Hottentots, that the cattle from Xhosaland were stolen, and some Xhosa shot dead. To continue these same expeditions Bothma hired his wagon to Hendrik Frederik Prinsloo in exchange for a number of cattle and thus took part surreptitiously.

In 1786 the heemraden and military officers of Graaff-Reinet, including Adriaan van Jaarsveld, David de Villiers, D. S. van der Merwe, Petrus Burger, Josua Joubert and J. G. Tregard—all prominent frontier farmers—declared that it was a well-known fact "that in the latest disputes the inhabitants have had with the Xhosa, and which certainly have been caused by some wanton and greedy persons, the Xhosa during the commotion have carried away much livestock belonging to the peaceable inhabitants, as well as some from those who had offended them."[39]

In the same year Woeke, the newly appointed landdrost of Graaff-Reinet, sent a report to the governor regarding the state of affairs on the frontier. He was afraid, he wrote, that the peace with the Xhosa would not be of long duration. In Agter-Bruintjieshoogte a Xhosa was shot dead trying to steal a sheep. And now, Woeke pointed out to the governor,

> that this incident with the Kaffer is similar to that which took place in the year 17——, and it clearly seems to me, the undersigned, that it was caused by Willem Prinsloo, who under the pretext that the Xhosa had stolen a sheep from him, shot one of them dead, whereupon the Xhosa rose up and attacked the

inhabitants, resulting in a terrible slaughter of the Xhosa and the ruin of many inhabitants.[40]

We can now summarize these fragmentary facts, which throw light on the causes of the first clash with the Bantu, as follows: it was generally believed by contemporaries that the first open hostilities, occurring toward the end of 1779 beyond Bruintjieshoogte between the frontier farmers and the Xhosa, were caused by the reprehensible behavior of a few irresponsible frontier farmers. It further appears as if the cattle thievery by the Xhosa did not play an important role in causing that conflict. No evidence could be found that the Xhosa stole any livestock from the farmers beyond Bruintjieshoogte before the end of 1779, or that they themselves were guilty of committing any other acts of violence. And what is still more important: in all the references made by frontier farmers and other prominent persons in the colony regarding the origin of the troubles, livestock theft or the use of violence by the Xhosa is never mentioned.

There was noticeable friction between the farmers and the Xhosa in the Swellendam district, more specifically along the Bushmans River, during the last half of 1779. The nature of those disturbances is not altogether clear. There was, however, a strained relationship that probably arose from the colonists' discontent over the Xhosa presence west of the Fish River and their unwillingness to depart, at the farmers' insistence, as they had promised.

The farmers' flight from the Bushmans River was probably caused by fear of the Xhosa rather than by any acts of violence that were actually perpetrated. On the basis of the declarations from some farmers in the vicinity it appears as if there did occur a few livestock thefts along the Bushmans River but apparently these were minor incidents, and probably were related to a dispute between the Xhosa and Khoikhoi. These stock thefts, however, were clearly only of local importance and could not be connected in any causal relationship with the conflict between the farmers and Xhosa beyond Bruintjieshoogte because other reasons were given by all witnesses to this conflict.

After open hostilities had broken out, there occurred a mutual seizure of livestock, but when it actually commenced is difficult to determine. Regrettably, we do not know when the Xhosa murder and the Prinsloo's expedition, which would have given rise to the Xhosa revolt, took place.

Whatever the incidental occurrences were, however, that formed the immediate inducement for the outbreak of hostilities, they were secondary to the primary cause. Given the hostile atmosphere already present, they could have been decisive in causing the inevitable to happen at that specific point in time. But the primary cause of the conflict with the Bantu on the eastern border undoubtedly lay in factors of a more general nature that had already caused dissatisfaction for several years. The farmers found the Xhosa presence west of the Fish River bothersome. With large herds of cattle that were grazed communally, they destroyed the farmers' pastures. In addition, they left significant numbers of isolated frontier farmers feeling insecure. Some had abandoned their farms out of fear of the Xhosa and those that remained behind felt that either that the Xhosa must be expelled or the farmers would be obliged finally to vacate the grazing lands west of the Fish River. Understandably, the farmers began to grow impatient when the Xhosa, in spite of their promises to van Plettenberg, remained encamped to the west of the Fish River. And in these circumstances some frontier farmers could have employed reprehensible means to make the Xhosa leave.

In this connection landdrost De Wet's commentary on Joubert's accusation against the Prinsloos is very interesting:

> But it is meanwhile certain as well that the family of Willem Prinsloo, Sr., for the most part are harmful and agitating inhabitants of that country, who will leave nothing undone in attempting all that is possible to force the Xhosa to dislodge from there, with a view to expanding the size of their own fields. Thus the promise made to Your Excellency by the Xhosa Chief Koba that, together with his people, he would make his way across the Great Fish River, is already taken as a good pretext for justifying their violently forcing the Xhosa to fulfill this promise (although there have been no orders given as yet for such extreme measures to be taken.)[41]

Van Plettenberg offers a similar interpretation of the conflict on the eastern frontier. According to him it was also principally the result of the conflict of interest over grazing land. When the reports regarding the Frontier War were discussed in the Council of Policy, the governor referred to the agreements that he had made with the Xhosa chiefs on his journey of 1778:

> in order to avoid all disputes with our inhabitants living thereabout, the Fish River should serve as a boundary between them, and consequently the above-mentioned Xhosa with their livestock should not have the freedom to travel hither over the said river, where our inhabitants have their fields, and on the other hand our people should stay on this side of that river.

The Xhosa had, "contrary to the solemn promise made to Your Excellency, nevertheless moved over to this side of the Fish River, and with their livestock were even settled on our inhabitants' farms, and thereby the troubles that have arisen with the said Xhosa, principally originated."[42]

The Further Course of the Conflict

The two commandos that marched out during the summer of 1779–80 against the Xhosa had no success at reestablishing the peace. In the winter of 1780 Rharhabe, the Xhosa paramount chief, returned a large number of the cattle the Xhosa had stolen from the farmers in the war, under the supervision of ten of his men. Along the way, however, the majority of the livestock were taken away by some subchiefs who had separated themselves from Rharhabe and settled between the Koonap and the Kat rivers.

Rharhabe was desirous of concluding peace, but first his rebellious subjects had to be brought under submission. He thus requested that van Jaarsveld come to an appointed place along the Koonap River and, once there, to have him sent for so that they could discuss matters. If the rebellious chiefs were unwilling to give back the stolen livestock and to submit to their paramount chief, then van Jaarsveld must assault them from the front, while Rharhabe would attack from behind.

This message made van Jaarsveld entertain the hope that he would be able to make peace without having to begin hostilities anew. He departed with a commando of 124 men and sent a message summoning Rharhabe, but the Xhosa chief did not turn up. Van Jaarsveld himself supposed that perhaps the message did not get through, but his men were afraid of treachery. Van Jaarsveld then had to disband the commando and turn home without achieving his objective.[43]

Toward the middle of 1780 most of the Xhosa were apparently east of the Fish, but as long as no peace was concluded, the farmers were afraid to take possession again of the abandoned fields. This caused a critical state of affairs. Reported van Jaarsveld,

> With the flight away from the border the people settled so closely pressed together that our cattle were all skin and bones and were dying daily from hunger. And if the deserted fields are no longer able to be inhabited, there shall be no more cattle to sell within a year, that is to say here in the district.[44]

He therefore asked the landdrost and military council to give him orders for reestablishing the peace. He would then make an effort to have the stolen livestock returned and to persuade the chiefs living between the Kat and the Koonap rivers to make peace with Rharhabe.

Some months elapsed before van Jaarsveld was appointed as commandant of the eastern frontier districts and received orders to make peace with the Xhosa.[45] In the meantime the Xhosa had again entered the colony. A few of the lesser chiefs went with their men to encamp in the Assegaai Bush. Langa came over the Bushmans River, evidently without any hostile intentions, and as a sign of peace sent back 101 cattle and nine horses that his people had stolen from the farmers. At the same time he asked permission to remain for three days with his livestock west of the Bushmans River. His request was granted, and then he evidently supposed it was now safe to remain in the colony, for when the three days were up he did not leave.

In these circumstances the Swellendam farmers became disgruntled and restless and Field-Cornet Bauer sent someone to Swellendam in the name of the farmers to petition that a commando be sent out in order to drive the Xhosa away.[46] In the Stellenbosch district as well, the Xhosa had crossed over the border and came to settle on the Kommadagga, "whereupon I assure Your Honor," wrote van Jaarsveld,

> that it shall shortly follow that the entire Bruintjieshoogte, Swartruggens, and Camdeboo shall be overwhelmed by them because among us all we have neither enough powder nor lead to defend ourselves. Then the Lord knows how miserable it shall go with our wives and children because the Xhosa that lay before us are the same we have already met in battle.[47]

Adriaan van Jaarsveld was appointed by the governor and Council of Policy on 24 October 1780 as commander of the eastern border districts with the rank of lieutenant. At the same time the landdrost, heemraden, and military officers were directed to draw up instructions for him and to present them before the Council of Policy for approval. On 5 December these instructions were endorsed by the heemraden at their meeting in Stellenbosch and on 27 December the Council of Policy approved them. Accordingly, van Jaarsveld was directed to put into practice the policy of territorial segregation that van Plettenberg had prescribed in 1778 and that was ratified once again on 14 November 1780 by a resolution of the council.[48]

"With the Xhosa, who are themselves a peaceable and timid nation, you must try to arrive at a permanent and lasting treaty of peace," read the instruction. This treaty had to be concluded with the Xhosa chiefs, who, in their own names and that of all their subjects had to

> promise and bind themselves, to be in abeyance of the regulations already drawn up by His Excellency the governor during his journey two years ago. To let the Great Fish River serve as a boundary between the territory of the Honorable Company and their own, so that they shall never have the power to make any claims on the lands and fields on this side, nor also that any use shall be, nor be allowed to be, made by our inhabitants on the lands and grazing areas on the other side of the aforementioned Fish River.[49]

If the Xhosa were unwilling to follow van Plettenberg's border regulations, "and were also not willing to let themselves be persuaded thereto by reason, a considerable and well-armed commando must be assembled at once and thus forcibly compel them to make for and to remain on the other side of the Great Fish River."[50]

The Commando of 1781

Such a peace treaty was evidently entered into, but the Xhosa did not hold themselves to it for very long. This is clear from a report that Adriaan van Jaarsveld forwarded to the military council on 20 July 1781 from Camdeboo. "Since the recently concluded treaty with the Xhosa, they have again moved in among our inhabitants with all their belongings, so that it became of the utmost necessity that resistance should be offered to the potential violence that daily threatened us. Thus I have assembled a strong commando, and begun to drive off the Xhosa,"* begins this interesting report, that clearly identifies the root cause for the problems on the eastern frontier.

On 23 May van Jaarsveld ordered Chief Koba to leave the colony with his followers and also to warn the other chiefs to again return to their own land. Koba was sensible enough not to answer back. He departed and van Jaarsveld went with his commando farther up into Agter-Bruintjieshoogte,

> where I closely examined all the Xhosa messages as well as the disturbances they had committed anew at night on the farms, with their occupying the farms and taking away by force the faithful servants from the farms of the inhabitants. I found the matter had reached a critical point, and it was of the utmost urgency that it be stopped either by peaceful means or with force.

On 1 June, in the name of the governor, the commander ordered three other chiefs to leave, and returned in peace with his

---

*[Trans. note: This and the following quotations are taken from the report cited in n. 51 below.]

commando to his camp. The following day, however, he encountered the chiefs again on the same spot and they then declared frankly that they would not leave. Just as on the previous day, the Xhosa again took up positions among the burghers with weapons in their hands. The commander also noticed that he was continually surrounded by ten or twelve armed Xhosa, but he thought they came to stand close by him in order to be able to hear better what he had to say. His interpreter warned him secretly, however, that he not only knew the wily ways of the Xhosa but had also heard them encourage each other to push in among the farmers, ostensibly to ask for tobacco, because then the burghers could not fire on them. Van Jaarsveld took the words of his interpreter to heart and for the last time ordered the Xhosa to leave. At the same time he warned them he would shoot if they were not gone in two days.

Van Jaarsveld visited the Xhosa for the third time on 6 June. Although he attempted to prevent it, they again mingled among the burghers with their weapons in hand. The commander was afraid that he would lose many men if the Xhosa attacked first, and he decided then and there to beat them to it. He drew up the commando in a line so that the burghers could fire to the front and to the rear. He had all the tobacco that was to be found brought together and cut up into small pieces. Then he threw the tobacco on the ground about twelve feet in front of the commando, and the Xhosa were invited to pick it up. "At this they ran out from among us, having forgotten their plan, at which I gave the command to fire, whereupon the three aforementioned chiefs were taken by surprise and slain together with all their armed men, and a portion of their cattle, numbering eight hundred, taken away."

On 10 June the commando departed from the farm of Willem Prinsloo, where they had camped up to now, toward the Bosberg. Here van Jaarsveld came upon Koba and two other captains. They were, as would appear from the commander's report, attacked without any preliminary formalities and 1,030 cattle were captured. After that van Jaarsveld informed the Xhosa through a captive that he was prepared to make peace and proceed to a mutual exchange of cattle if they would vacate the ground to the west of the Fish River in keeping with their promises to van Plettenberg. The Xhosa did not react favorably to this proposal. In the evening, while the commando was busy preparing to go back, the Xhosa appeared in great numbers out of the bushes and called out to the farmers that they would get their cattle back again as soon as it became dark. At this a number of the burghers separated from the commando and drove the Xhosa back during the night with a loss of six of their men.

A few days later the commando encountered chief Langa, who had remained along the Bushmans River. Van Jaarsveld "beseeched him in the name of the governor, for the preservation of the treaty of peace that had been concluded, to return again peacefully to his own land and henceforth never again to make any claims to the lands on this side of the said river, to which he consented and complied." The following day van Jaarsveld planned to order two other chiefs in the vicinity to leave, but when he came to their campsite he found they had fled from there on the previous night— sensibly enough.

Van Jaarsveld again came across the three chiefs that attacked the commando beyond the Bosberg, along the Baviaans River. They stood on the mountains and called to the farmers that they would resume the battle, would exchange no cattle, and would get their cattle back again. Van Jaarsveld then decided to wait a little while, and on 19 June the commando was disbanded for three weeks. When the commando—ninety-two "Christians" and forty Khoi-khoi strong—reunited again on 9 July, the Xhosa had not yet attacked. They had, however, set out for the Fish River and along the way had chased off the cattle of two farmers and severely wounded the herdsmen. On 16 July the commando came across the chief Taathoe, whom they had previously attacked behind the Bosberg, at the confluence of the Great and Little Fish rivers, "defeated him with all his warriors," and captured 1,500 cattle. On the burghers' side only three Khoikhoi were injured. The following day the commando encountered Thiete and Zeka, the two chiefs who had fled from the Bushmans River before van Jaarsveld could deal with them, to the west of the Fish and they "attacked, but because of the abundance of bushes could kill very few." The commando, nevertheless, succeeded in taking 2,000 cattle.

After nearly two months in the field, the commando was disbanded. Twelve men were left behind to protect the outposts and the corporals and field-cornets were given precise instructions on how to act toward the Xhosa. The commandants were very satisfied with the success of their undertaking, considering that they all felt "that the still plundering Bushmen cannot be properly beaten unless the rebellious Xhosa are first driven back with force and made to stay in their own country."[51]

Van Jaarsveld's instructions had prescribed that a mutual exchange of "each other's cattle that were seized by force, both by hostile actions and otherwise," must occur, if he was to succeed in making peace with the Xhosa. He had to take care that none of the Xhosa livestock remained in the farmers' hands, and consequently everyone that shared in the spoils of Joubert and Ferairre's commando had to return as many cattle as they had received at the time.[52] If, however, the Xhosa refused to cross over the Fish River voluntarily and had to be driven across with force, "it is not expected that a mutual exchange of each other's seized livestock can come to pass."[53] In such a case van Jaarsveld would not be allowed, however, to seize any more livestock from the Xhosa, unless the commando discovered cattle belonging to the farmers among the Xhosa herds. Such livestock should then be taken back and restored to their lawful owners.[54]

According to van Jaarsveld's report he raised the question of livestock exchange for discussion, but evidently this did not interest the Xhosa. Then, in keeping with the spirit of his instructions, after having expelled them by force the commandant dropped the question. Furthermore, in contravention of his instructions, he took 5,300 cattle from the Xhosa. Among these, however, were a number of cattle that belonged to the farmers. With the disbanding of the commando 10 percent of the captured cattle were divided equally among the burghers who had personally taken part in the commando from beginning to end. The remainder were divided among the farmers who had lost livestock during the war, on the basis of "the number that they have declared in good conscience." The farmers got back enough breeding cattle to make good their stated losses, if the fifty-three cattle per person that were shared out following Joubert's commando are included. But as to the draft oxen, the farmers received only about 43 percent of the total they had declared. The commander kept exact records of the livestock divided out, so that the animals could be taken back again in case the government did not approve the distribution.[55]

For the contravention of his instructions van Jaarsveld offered the following reasons: in most cases the cattle that the Xhosa stole from the farmers could no longer be identified because the ears and tails were cut off. Poverty and starvation threatened the farmers that had been robbed of their means of livelihood, and the cattle confiscated from the Xhosa did not even make good the farmers' losses. The final reason offered was "to compel the Xhosa by the loss of their cattle to remain in their own country in future."[56] On receipt of van Jaarsveld's report, the Military Council of Stellenbosch studied it thoroughly, decided that the commander acted in accordance with his instructions, and asked the governor to approve of everything that occurred on the commando.[57] On 18 October the Council of Policy complied with this request,[58] but on 6 November the following decision was taken in reference to the distribution of the cattle that were seized:

> It is understood that the divided cattle may be retained for this time, without, however, any conclusions being drawn from this for the future, much less making the same serve as a basis on which any quarrel by our inhabitants against the above-mentioned Xhosa must be sought, and thus more be stolen from them.[59]

After the commando of 1781 we hear nothing from the Xhosa for a long time. Evidently they remained to the east of the Fish River and kept quiet. On 8 May 1782 Adriaan van Jaarsveld wrote to the Military Council of Stellenbosch,". . . the Xhosa have come around to make an offer of peace, on the condition that we, as a token of the same, must carry out a mutual exchange of the stolen cattle, as if this was their share."[60] It does not appear, however, that van Jaarsveld gave any notice to the proposal, and evidently after the Xhosa were driven across the Fish River, another formal treaty of peace was never concluded with them.

From the foregoing representation of the first encounter between white and Bantu, it would appear that the farmers from 1778 until 1781 were not confronted with bandits that invaded the colony and perpetrated all sorts of acts of violence, but with rivals over grazing land, who could not be moved by fine words to vacate the country to the west of the Fish River and who were finally driven away by force of arms.

The Second Frontier War

It was evidently not long after the first encounter on the eastern frontier that hostile relations between white and Bantu were renewed. The farmers were not at all afraid to visit Xhosaland,[61] and some even requested farms on the eastern side of the Fish River.[62] The Xhosa as well did not long remain away from the colony. From an instruction to Field-Commandant D. W. Kühne, dated 21 November 1781,[63] it appears as if Xhosa were already west of the Fish by this time. The Cape government, however, was seemingly more concerned about the frontier farmers' inclination to migrate as about Xhosa border violations, and did nothing to restrict the latter to the east of the Fish River.

In 1785 the governor and the Council of Policy—partly out of fear that the farmers' migration across the Fish River could give rise to new problems with the Bantu—decided to establish a drostdy in Camdeboo.[64] The landdrost and heemraden of the new district, formed from the eastern parts of the Stellenbosch and Swellendam districts, received jurisdiction over the entire Xhosa frontier.[65] To prevent a second clash between white and Bantu, the landdrost had to keep the farmers out of Xhosaland and those already settled beyond the border were to return immediately to the colony.[66] Moreover, despite numerous ordinances, the unlawful cattle trade with the Xhosa was rapidly beginning to flourish again,[67] and again was forbidden in the strongest terms.[68]

Shortly after his arrival in Graaff-Reinet, Woeke conferred with the nearest Xhosa chiefs and presents were exchanged as a sign of friendship. The landdrost doubted, however, if this friendship could be of long duration. In August 1786 a Xhosa was shot dead trying to steal a sheep beyond Bruintjieshoogte. A few farmers went into Xhosaland with their livestock, while others, under the pretense of shooting wild game, migrated across the border with their wagons. At the mouth of the Fish River the unlawful livestock trade was carried on more vigorously than ever before. Xhosa had settled among the farmers everywhere and streamed into the colony daily to offer their cattle for barter. Even in the Swellendam district, which now extended no further eastward than the Gamtoos River after the formation of the new Graaff-Reinet district, there were Xhosa wandering about.[69]

Woeke recognized that only a strong and resolute policy could save the position on the eastern frontier. He was inclined to use force to expel the Xhosa to their own land if they did not want to return voluntarily. He also wanted to destroy the unlawful livestock trade root and branch. But to maintain public order on the eastern frontier, where everyone did as they wanted, he had to have more power. He repeatedly asked, therefore, that fifty to sixty soldiers be stationed along the Fish River to see that the laws were obeyed. Objections were raised against this proposal, however, that it would give the new landdrost too much power and prestige.[70] With his hands tied, bitterly disappointed, Woeke had to look on "as destruction proceeded everywhere in the country and is to be found in the most distant corners, and thus, if it is not suppressed, the problem will increase and each person will follow his own will indiscriminately and be able to do anything he wants." He bitterly declared that it "is also not proper to appoint regents to whom the power is not given to be able to keep the transgressors in check."[71]

Unfortunately the central government was not prepared to support Woeke. It evidently expected that Woeke must rescue the situation with warnings and fine words. The farmers had to be shown therefore that they not only exposed themselves to danger through the livestock trade with the Xhosa, but also gave the Xhosa a reason to bring their livestock west of the Fish River. The Xhosa had to be convinced that it could have fatal consequences for them, as experience had taught, if they did not remain in their own land. And the chiefs had to be persuaded to punish their subjects who invaded the colony.[72] The border situation thus remained much as Woeke found it when he first arrived in Graaff-Reinet.

Apparently all went well for a while, but toward the end of 1788 complaints began to come in again about Xhosa behavior. While Woeke was absent from the drostdy, the secretary, Wägener, received a petition from nine farmers who lived in the Zuurveld beyond Bruintjieshoogte. In this petition they, in a "fairly exalted tone, sought assistance from the Landdrost and Military Council against the Xhosa nation. According to them, a confrontation would begin soon with this nation, and some of the inhabitants have long wished for such a fight so as, if possible, to seize the Xhosa cattle, which they have always regarded with envy."[73]

Wägener forwarded a copy of the petition—which is unfortunately nowhere to be found—to the governor and explained that he did not want to convene a special session of the military council, "as the field-cornets named above have yet to report on the hostilities which the petitioners profess to be perpetrated by the Xhosa nation, and also because the persons named in their petition, and from whom the aforesaid nation stole cattle, have themselves not yet complained about this problem."[74]

The nine burghers who had drawn up the petition, received the following answer from Wägener:

> I find that there is not enough satisfactory evidence furnished in this petition concerning the molestations perpetrated by the Xhosa nation, as well as the cattle they are alleged to have robbed from Jan Meijer, Andries du Preez, Coenraad de Buijs, and Jacobus Steijn, in order, according to your wish, to send aid and assistance, and attack that nation. It appears very doubtful, because, though it might be true regarding the nation beginning a confrontation, the respective militia officers make no report of it, and the above-mentioned persons, from whom the cattle supposedly have been stolen, have also made no complaint about this and have not signed the petition.[75]

It is impossible to determine the significance of this petition. Wägener doubted, evidently not without reason, the credibility of the stories reported in it. One thing is clear in any case, namely that the farmers would have been pleased if the Xhosa were driven away. In this specific case it cannot be established whether it was because the farmers lived with the thievery of the Xhosa, because they experienced acts of violence or feared them, or because the Xhosa cattle devoured their pasturage. Discontent over the Xhosa presence west of the Fish River, however, grew daily and vague accusations about their behavior increased.

On 9 February 1789 Woeke reported as follows to the governor,

> From various incoming reports and inhabitants' complaints it appears very clear and plainly obvious, that the Xhosa, includ-

ing four chiefs on this side of the Great Fish River, have settled
with a great number of their followers among the inhabitants,
and roam about all over the colony by the tens and twenties,
committing all sorts of mischief and molestations, stealing livestock and in general not willing to listen to a friendly warning,
and up to no good.

Under these circumstances he commanded Captain Kühne
with three field-cornets and their subordinates

once again to try to drive the Xhosa across the Great Fish River,
where they shall remain. If not, we shall be obliged to deal
with them in a hostile manner and bring them to reason. This
undertaking, however, shall not come to pass before it becomes
absolutely necessary, at which time I, together with the deputy
heemraden and military officials, shall meet with the chief, and
before attacking yet see if they will listen to a friendly warning.

Before an eventual attack on the Xhosa was made, further care was
to be taken that the frontier farmers retire back in time, and an
armed force would be placed along the Fish River in order to prevent a general invasion of the colony.

Woeke realized it would be impossible to maintain a lasting
peace on the frontier without stationing a permanent military force
along the Fish River, and therefore he repeated his request for sixty
to one hundred soldiers. He doubted, however, whether the farmers
would tolerate the wantonness and thievery of the Xhosa until the
soldiers arrived. "All the inhabitants' reports and charges regarding
the Xhosa are valid," he explained. "It is painful to allow oneself
to be robbed of one's livestock and possessions without reprisals,
which with this nation can only be done by force of arms." Finally,
he asked for a speedy answer, so that, if it was still possible, the
seemingly imminent war with the Xhosa could be prevented.[76]

Before Woeke's correspondence reached Cape Town, Captain
Kühne[77] had carried out his instructions.[78] He had driven the
Xhosa from the Bushmans River up to the Great Fish River without
much difficulty, but unfortunately the latter river was so full no
one could cross over it. As a result, Kühne had to allow the Xhosa
to temporarily remain west of the Fish River, but he promised to
drive them into Xhosaland as soon as the stream was passable. One

chief absolutely refused to cross the Fish River. He maintained that he had purchased the land between the Kowie and the Fish River, and it was thus his legal property. Furthermore, reported Kühne, "No hostilities had begun as yet, but the Xhosa continued to commit general annoyances and had stabbed to death a horse belonging to the burgher Jan Viljoen." What would happen in the future he could not say because many farmers had Xhosa in their service. He therefore requested further instructions from Woeke.[79]

Perhaps in this connection it is interesting to refer to the report from a traveler who was riding through Xhosaland at that time. He tells of the Xhosa's riches in cattle and declares,

> on my return journey, along the Karega River, sixteen hours this side of the Great Fish River, there were several thousand Xhosa who had roughly more than 16,000 head of horned cattle. They had quite surrounded the homestead of a certain Piet Lombards, and although they had come here with friendly intentions merely to hunt, their enormous herds still stripped the colonists' grazing lands of precious fodder. At times, moreover, they make demands on the inhabitants of these farms and it was often quite risky to resist the cheeky claims of a crowd of armed cattle herders. Some time after that they were urged by threats from a hostile commando, to withdraw again back into their own land. Yet just then the Great Fish River was swollen and rising, so had they complied with the residents' wishes, a large number of Xhosa would have perished together with many cattle.[80]

After Woeke had received the report of Kühne's activities, he decided not to take any further action before receiving instructions from the Council of Policy. Meanwhile he instituted inquiries about the behavior of the Xhosa and ordered the farmers, by means of the field cornets, to release the Xhosa they had in service. He wrote to the governor that he planned to wait until the Fish River had subsided and then see if the Xhosa would voluntarily return to their land and remain there, or whether they would have to be compelled to do so. If all dealings with the Xhosa were not halted, hostilities with them would eventually begin again and it could have fatal consequences for the colony. The only way in which to maintain a lasting peace on the frontier—he declared once again at the end—was to station soldiers there.[81]

On 20 March Woeke's first report on the strained relations between the farmers and the Xhosa was discussed in the Council of Policy.[82] The landdrost was warned "not to take premature, but well-considered steps, to use every means and to leave nothing untried that generally can maintain and promote peace and quiet between the inhabitants and the Xhosa." He must further investigate whether the Xhosa presence in the colony was not attributable to the supposed claims of specific chiefs to land west of the Fish River. (Woeke's letter of 3 March had not yet reached the governor at this stage, but the latter had apparently already learned of the Xhosa claims on the Zuurveld from other sources.) If this was truly the case, he had to try and

> exchange the presumed claims for all time for some trinkets and thereby compel these Xhosa to remove themselves from this to the other side of the Fish River and to stay there, in order that the inhabitants might be freed from annoyances, peace and quiet between both sides be preserved and the many harmful results that were feared be averted.

On receipt of these instructions Woeke called a joint meeting of the heemraden and military officers. They came together on 13 May. The landdrost pointed out to the meeting that the governor's assumption with reference to the Xhosa claim on the Zuurveld was not unwarranted. Captain Kühne, as well as militia officer Lucas Meyer, reported to him that the Xhosa chief Kaulta had laid a claim to the Zuurveld. Furthermore, the landdrost informed them that he had lately received reports and charges from militia officers as well as other burghers

> describing the annoyances and mischief committed against the inhabitants residing on this side of the Great Fish River by the Xhosa there. A multitude of Xhosa were also in this colony, wandering about in bands, and also committing insufferable annoyances. As a result the inhabitants, not individually but in groups and nearly unanimously, were saying that the Xhosa must be attacked and driven off with force. The landdrost had perceived these sentiments and observed that not only was a

war with that nation unavoidable, but also that many inhabitants appeared very favorably disposed toward it.[83]

Although the reports and accusations that came in regarding the conduct of the Xhosa were sometimes in direct contradiction, the situation on the eastern frontier in any case demanded serious attention and there was a feeling that something had to be done immediately. It was then decided to delegate a commission to travel to the eastern frontier. The commission had to attempt to buy out the Xhosa claims on the Zuurveld and in this manner compel them to cross the Fish River and to remain there. "Some trinkets" were to be distributed to the chiefs as tokens of peace and friendship, and the commission had to try and persuade them to keep their followers, who wandered about within the colony in bands, in Xhosaland. As an additional means to cut off communication between the farmers and the Xhosa, the ordinance of 19 July 1786, in which the livestock trade with the Xhosa and migration across the colonial borders was forbidden for the umpteenth time, had to be read out to the frontier farmers.[84]

Woeke, who had already on a previous occasion expressed the wish to negotiate personally with the Xhosa chiefs if Kühne's mission did not succeed, took upon himself the leadership of the commission. Accompanying Woeke as commission members were: the Heemraden Josua Joubert and Jacobus Gustaaff Tregard, the Military Council members Captain A. P. Burger and Captain Jan du Plessies, and also J. J. F. Wägener, the recently retired secretary of the district, and H. Maynier, the new secretary.

The commission[85] began its journey on 7 June 1789 but had to stop at the Kowie River, because the Fish River was swollen, and therefore the members could not reach Xhosaland. Khoikhoi messengers were then sent with presents to Chief Tshaka in order to inform him that the commission wanted to negotiate with him in order "to settle peacefully between him and the inhabitants their formal differences." Tshaka and Chungwa were prevented by illness from traveling, but they sent emissaries—among whom, a brother of Tshaka and a subordinate chief—to enter into negotiations with the commission in their name. Furthermore, Tshaka thanked them for the presents and as a reciprocal expression of friendship sent two slaughter oxen to the "great boss."

At the conference the landdrost asked the Xhosa

why they continued to remain on this and not the other side of the Great Fish River; that this river had been agreed to as the stipulated limit for the boundary.[86] The answer given by the Xhosa Prince in fairly understandable Low Dutch was that the Xhosa had already exchanged the tracts of land lying between the Kowie and Great Fish rivers for cattle with a certain Gonas Khoikhoi chief by the name of Ruiter many years before. Also, that the land which the Xhosa nation now inhabited on the other side of the afore-mentioned river was old and devoid of wild game, and they not only do not have enough grazing for their livestock, but are also deprived of their indispensable hides as *karosses* [skin cloaks] or covers.

Thereupon the commission pointed out that the Xhosa could not deny that the specific region legally belonged to the Company, that different Xhosa chiefs recognized the Fish River as the colony's boundary, and with regard to this point there existed absolutely no doubt. The Xhosa could not argue with this, but they answered

that without that land, for the reasons previously cited, they could not exist. They also well knew, that that land legally belonged to the Honorable Company; yet that their respective chiefs were favorably disposed to pay the Honorable Company for that land, but they could not do the same for the inhabitants.

Understandably, the commission declined this offer. The Xhosa were ordered to give their respective chiefs a complete account of the negotiations "and to warn the same to remove themselves to the other side of the above-mentioned Great Fish River and henceforth never again to settle with their livestock on this side of the often-mentioned river." The commission made clear that if they did this they could count on the colony's continued friendship. To furnish proof in anticipation of this a second consignment of presents was sent to the Xhosa chiefs.

After this meeting the commission began its return journey. The landdrost had still not personally set eyes on the Xhosa chiefs, but he had done everything that was possible under the circumstances in order to promote peace and friendship with the Xhosa, and the domestic affairs of the district called him home again to the drostdy. On the farm of Nicolaas Goede the landdrost found a large number of farmers gathered together who, as previously agreed, had been waiting for him there. He made known the contents of the ordinance of 19 July 1786, that was directed against the unlawful livestock trade with the Xhosa and migration across the border, and warned them to take seriously the regulations contained therein, to avoid all dealings with the Xhosa and to release the Xhosa that were in their service.

At this meeting a letter received by Burgher-Adjutant Barend Lindeque from his father was brought to the commission's attention. The letter was dated 25 June 1789 and splendidly restated the Xhosa position. In it was made known

> that all was not really in order with the Xhosa. They are on our farms and, having waited on the side until we were gone away, they now say that the Landdrost is gone, and what they are going to do, and that they only pretended to leave, and to nephew Campfer and his people they have said if they must pull out of the country than the Christians must get out of there as well.

He wanted to speak with the Xhosa, but they had refused to come; ". . . the Xhosa also said that Roelof had plotted with the Landdrost, that he fetched the Landdrost to chase their people away."[87]

This interesting letter was interpreted by the commission as follows:

> that there was something going on within the Xhosa nation, which showed that they suspected, not without reason, that the peace between their nation and the inhabitants could not last for a very long time and that the Xhosa did their utmost to block all efforts in order that they could keep their homes on this side of the Great Fish River and thus usurp a part of the legal property of the Company and arrogate to themselves the right as masters and rulers thereof. This appeared even more evident from the verbally transmitted incoming reports, which implied that the Xhosa had warned the Hottentot servants of some of the inhabitants, that the residents of that region of land would be the first to have to move away.[88]

The commission did not have time to tarry any longer on the frontier, but at the same time it felt that under these circumstances it could not permit matters to go their own way. It also did not consider it advisable to charge one of the frontier farmers who was personally involved in the Xhosa question with the responsibility of preventing any trouble. Accordingly, J. J. F. Wägener was left behind on the frontier to act in the name of the landdrost. He would be answerable to the landdrost and heemraden at Graaff-Reinet as well as to the governor in the Cape, but was authorized to take any steps he thought necessary. At the same time the frontier farmers were notified that they were to execute promptly and without any back talk take any order that Wägener gave them in the commission's name.[89]

Wägener did not wait long before sending on his first report together with the letters he had received to the landdrost and heemraden. On 3 July he had received a letter from militia officer Lucas Meyer containing news that the Xhosa still remained west of the Fish River, daily entered the colony, and had no plans to leave. The farmers that lived along the border had already begun to migrate back toward the colony "on account of the Xhosa, who were doing all manner of indecent things to the Christians." A Khoikhoi and a Xhosa from Langa's kraal had stolen cattle from Meyer.[90]

Wägener, who was obviously inclined not to let any news go unreported, related further, "that a Xhosa at the home of Roelof Kampfer had snatched away by the arm, in a brutal manner, a child of said Kampfer that sat by a fire, and occupied that child's place himself."[91] On 4 July he received a letter—signed by J. H. Viljoen, Roelof Campfer, and Roelof Janse van Vuuren—containing news that the Xhosa were on the move, but that several of their encampments still remained behind, "and these people had already loudly said to Kampfer three times that we must depart or they would attack us." The Xhosa had not yet attacked, but the farmers did not know when they might. The Xhosa walked around daily in bands, "and we intend to pull out because here there is so much death among our livestock."[92]

In order to prevent panic from developing on the frontier, Wägener forbade the farmers from taking flight on threat of severe punishment. In addition he gave orders to the field-cornets Lucas Meyer and Cornelis van Rooyen to instruct in his name the farmers living the furthest away, and whose farms were exposed to the most danger, to move together with three or four other families. This would enable them to protect their livestock, and to defend themselves if the Xhosa unexpectedly attacked.[93]

Wägener further reported according to a story from one of Langa's people, that Langa and Tshaka, who had been enemies for a long time, were once again friends. This story made him anxious and he accordingly made militia officer P. J. Delport aware of the story, warning him to be on his guard for Xhosa. Just like the other two militia officers, he was to have the people in his field-cornetcy move together and must warn them not to take flight on the basis of a mere rumor.[94]

With one voice, Wägener continues, the farmers assured him that if the Xhosa had evil intentions they would wait until there was no moonlight so that they could make their attack in the dark. In case it became necessary to fortify the frontier, he would conscript Cornelis van Rooyen's men between the Sundays and Gamtoos rivers. The majority of farmers were without ammunition, however, and he therefore asked the landdrost to send a supply. He hoped, in the meantime, to gain more information and also sent two emissaries to Tshaka.

At this point Wägener, who originally had volunteered to remain behind on the border, became very anxious to leave for the Cape. He asked that someone else be put in his place or that he be permitted to entrust his work there to Lucas Meyer. At the end he noted in a postscript that he had received a (secondhand) report that the Xhosa were advancing along the coast in large numbers. According to a thirdhand message originally spread by a Khoikhoi, the Xhosa were planning to attack and it was made clear to the colonists that the lot was cast first for captain Kühne and Lucas Meyer.[95]

The landdrost and heemraden of Graaff-Reinet were not very pleased with Wägener's work. They felt he had left them in the lurch, and ought to have known beforehand he had important business in the Cape and was only supplied with necessities. They were disappointed that the Xhosa were still west of the Fish River and by their behavior had created fear among the inhabitants. On the other hand, the majority of messages Wägener had received were nothing other than rumors spread by Khoikhoi and the persons concerned. Such communications were not reliable and Wägener had done nothing to establish their trustworthiness. Consequently the heemraad meeting decided that "from this one can make up any hostilities on the part of the Xhosa without sufficient proof."*

Wägener was warned to take care that no panic developed, and that no hostilities commenced that sprang from ungrounded fears. The landdrost could not send him ammunition, because there was barely enough powder and lead at the drostdy for their own defense. The shortage of powder and lead in the border districts could be ascribed in large measure to the farmers' carelessness. They were repeatedly warned by means of posters to make provision for shortages. Militia officers and farmers who were well supplied had to help the others so far as possible. Finally, Wägener was ordered to transfer his power to militia officer P. J. Delport and to give him the necessary instructions. Apparently Lucas Meyer, whom Wägener had recommended, was according to his own report a public enemy of the Xhosa and therefore unsuited for the execution of the proper orders.[96]

The Xhosa never made the long-awaited attack. Wägener was fortunately able to prevent the flight of the farmers and to fortify the border districts by allowing the farmers to move together. When the Xhosa saw the farmers take these steps, they were alarmed, fled head over heels across the Fish River and sought shelter in the thick bush. "By all these circumstances the distrust on the side of the Xhosa, as well as on that of the inhabitants, was even greater and everyone was on his guard and mindful of his safety."[97]

Wägener then let the Xhosa know through their own people that the farmers were capable of defending themselves in case the Xhosa attacked, but that they never intended to be the first to attack. A few days later he went with twelve men to discuss matters with Chief Tshaka. He entered into negotiations with Tshaka and Chungwa—the son of the old chief—but "not withstanding all the trouble he had gone to, he could not achieve peace with his countrymen on the other side of the Great Fish River. The situation, however, has been managed so far, so that at present all is in peace and quiet, and the nation's inhabitants have nothing to fear."[98]

---

*[Trans. note: No source is given for this quote in the Afrikaans edition.]

The account of these negotiations, which took place on 18 July 1789, casts an interesting light on the nature and origin of our border troubles with the Bantu. Chungwa again laid claim in his father's name to the region between the Fish and the Kowie rivers by virtue of the fact that it had been exchanged with the Khoikhoi chief Ruiter. Furthermore, he assured Wägener "that they could never or would never relinquish it—that the region was their life, and if they had to do without it they *would lose their life*."

Thereupon, Wägener pointed out to Chungwa that Ruiter had no right to sell the land, and thus had misled the Xhosa. The governor had sent him (Wägener) to the Xhosa to undeceive them in this regard and to make good the damage that was done them in the interest of mutual friendship. He expected, however, that when the compensation was paid, "the Xhosa nation would retire, and would never make any further claim on that land."

All of Wägener's efforts to get the Xhosa to renounce their claims to the Zuurveld appeared fruitless, however, since

> Coungoa (Chungwa) in guarded terms intimated he would rather die than give up the land. For this reason the undersigned found it advisable to come to a provisional accord as follows: that the Xhosa would be able to settle *provisionally* at their old kraals with their livestock. When the order of the *Great Chief* came, however, they then would have to move across the Great Fish River, and must be obedient to that order; that the Comp(any) would never ever renounce its claim to the region, but that in consideration of the mutual friendship and because of deaths among the Xhosa's cattle, they would be permitted to remain settled there yet a while.
>
> Upon which words the above-mentioned Chief Coungoa replied not at all rudely that he wanted to put them at ease, when the handing out of the land depended on the *Great Chief*, that he well knew he would never begrudge him [Chungwa] of the region. That he thus found his adversary only in the inhabitants living here in the neighborhood.[99]

After this arrangement, Wägener departed. It was very misguided of him to allow the Xhosa, who now were at last beyond the Fish River, to turn back toward the colony, albeit only provisionally. On the other hand, it was also clear he would not be able to hold the Xhosa east of the border with fine words and presents. And the central government was not yet prepared to provide him with stronger means to uphold his policy of territorial and economical segregation.

After Wägener had quit the frontier and the landdrost and heemraden had received his reports, the commission that he represented there considered their task concluded. They were sorry, the commission explained, that they were not able to amicably to persuade the Xhosa to renounce their supposed claims to the Zuurveld. "And knowing the peaceable sentiments of Your Honorable Sirs and not employing any forceable means to make that nation depart," they waited for further instructions.

The commission, however, warned the government that, with an eye on the future, it would not be advisable to concede too much to the Xhosa. For the sake of precedent it was necessary that the government adopt a definite position. It was not worth being at odds with the Xhosa for the sake of the region to which they made claim. Without doing very much harm to the Company, the government could renounce the land "as they did not have to be afraid that the other Xhosa Chiefs, because of this renunciation, would also come to make supposed claims and so by the ceding of this limited region to the inhabitants the way would be opened to a public exchange, which always in all cases must be very carefully guarded against."[100]

Thus, the storm that had threatened the frontier toward the middle of 1789, moved off following the agreement made with the Xhosa by Wägener. These unsatisfactory arrangements, with which the central government was evidently altogether content, likely prevented a possible clash between white and Bantu. On the fertile ground of mutual distrust hostilities could very easily have broken out if a government representative had not been present who could keep both parties in check in the border districts. But the permanent cause of the strife between the frontier farmers and the Xhosa was not yet eliminated. Wägener had barely left the border districts when the Xhosa again came across the Fish River. Apparently they did not have hostile intentions or perpetrate any hostilities; there are no complaints to that effect in the documents of the time and the only available report about them reads that they "are still at peace but some are settled on this side of the Great Fish River."[101]

The farmers, nevertheless, found the Xhosa presence in the colony very trying, as seems plain from the following note to Woeke from Gerrit Scheepers, Sr., dated 27 December 1789:

> This serves to register my complaint regarding the annoyance that I have from the Xhosa chief Langa on my farm called Sweet Milk Fountain. He is settled with his people and live-stock between me and Zwaanepoel and not only allows the fields to be depastured and the water supply drained but de-stroys the land with burning. Not only me but almost everyone there around Bushmans River has Xhosa around their people to their annoyance. They move about in force, so that I fear the land shall be occupied before men there are prepared or can be summoned. I have already been ruined by the Xhosa one time, My Lord, and fear for the second time, which is why I ask that My Lord will still care for us as a father because we suffer greatly on our quitrent farms from oppression by the heathens. I therefore petition a speedy answer.[102]

It was thus, toward the end of 1789, the same old story: the Xhosa come to settle with their livestock to the west of the Fish River and destroy the grazing land of the farmers. This makes the farmers disgruntled. What is more, the farmers feel themselves threatened by the increasing immigration of the Xhosa into the colony. These circumstances together would be a fertile ground upon which renewed friction with the Xhosa could develop.[103]

The representation that we have given above of the friction in 1789 between the farmers and the Xhosa differs on important points with Dr. Theal's description of "the second Kaffir invasion of the colony."[104]

Dr. Theal creates the impression that in March 1789 the Xhosa suddenly invaded the colony and that the farmers, who were not capable of defending themselves, escaped before the Xhosa on-slaught. From the documents of the time it would appear, however, that toward the end of the 1780s the Xhosa entered the colony little by little but without any hostile intentions.

Furthermore, Dr. Theal gives the impression that the purpose of Chief Kühne's expedition was to punish the Xhosa and to take

back stolen cattle, and he erroneously presents this as if the commando, when it stood on the point of plucking the fruits of an easy
victory, had to disband on instructions from the Cape to the bitter
disappointment of the burghers. In point of fact the intent was
quite obviously that Kühne must not use force. To be sure, Woeke
considered it *possible* to attack the Xhosa with force. But this would
not happen before the landdrost himself had conferred with the
chiefs and the necessary steps were taken for the protection of the
frontier farmers. In addition, from the report on Kühne's performance, it seems quite obvious that he himself had no intention, in
contravention of his instructions, of treating the Xhosa in a hostile
way or to take away their livestock. And he drove the Xhosa no
farther than the Fish River simply and solely because the river was
full and no one could cross over it. In this connection, moreover,
Dr. Theal commits a chronological error. Kühne's commando was
not disbanded at a critical moment on instructions from the governor. On 3 March 1789 Woeke reported that Kühne had already
executed his instructions and on 20 March the Council of Policy
learned for the first time of the disturbances on the border. Only
then did this council send Woeke instructions with regard to the
border question.

Finally, Dr. Theal relates that the farmers, when they had to
take flight before the Xhosa, were not able to save all of their
livestock (p. 178). He speaks vaguely as well of "the losses sustained
by the invasion" and of the "disastrous consequences (of the second
Xhosa invasion of the colony) to the farmers" (p. 181) and wants
to maintain that the farmers on the Xhosa border were continually
subjected to "ruinous losses" following Wägener's departure and
were compelled to stand guard over their livestock with loaded
weapons (p. 182).

We have not succeeded in finding in the documents of the time
the communications that mention the losses the farmers suffered
as a result of the "invasion" of the Xhosa. Even if the occasional
messages about alleged Xhosa cattle stealing, to which we have
referred above, were also absolutely trustworthy, we still cannot
simply assume on that basis that in 1789 stock thefts played an
important role in the troubles on the border. It is also very peculiar
that the detailed report of the commission, which had been set up
to investigate the border troubles, makes no mention of stock thefts.

For as long as the commission (and later its representative Wägener) was on the frontier it received various reports and complaints about the behavior of the Xhosa, but about stock thefts there are never any complaints. The commission, which was appointed in order to remove the causes of the disputes between white and Bantu in a peaceful manner, also never raised this question in all its negotiations with the Xhosa. And from the two reports the landdrost received after Wägener's departure, regarding the behavior of the Xhosa, one can also not infer that the frontier farmers at the time continually suffered from Xhosa thievery. If, however, this was the case, why then would the farmers, who would gladly see the Xhosa driven away, keep silent about it in communications that directly dealt with the Xhosa? Their chances of having the Xhosa driven out of the colony would have been so much better if they could have declared as well that the Xhosa had stolen their livestock.

One must not consider too quickly the *argumentum ex silentio* [argument from silence] as conclusive proof. But in this case it leaves one strongly in doubt about Dr. Theal's representation of the role that stock thefts played at the end of the 1780s on the eastern frontier.

On the other hand, it appears quite clear from various documents that conflicts of interest over land were a very important cause for the disputes and the strained relations that developed in 1789 between the farmers and the Xhosa on the border. And to this Dr. Theal does not give the necessary attention. The Xhosa were eager—for whatever reason—to possess the Zuurveld; they were disgruntled because the farmers had "schemed" with the landdrost to drive them away, and they looked upon the farmers—not the government—as their rivals in the Zuurveld. The farmers were again discontented about the Xhosa presence west of the Fish River and they wanted them chased out. Individual farmers found it frustrating to be oppressed "by heathens," who used and ruined their pastures, on their quitrent farms.[105] And the farmers as a group— even those who did not personally experience the Xhosa annoyances—regarded their presence in the colony as a continual threat. They felt the Xhosa had to be expelled if the farmers wanted to ensure possession of the Zuurveld for themselves. If the Xhosa were allowed to stream into the Zuurveld in large numbers, eventually the farmers would be compelled, as a result of real or feared violence from them, to evacuate the area. The Company also considered the taking of its territory as a serious threat and laid the most emphasis upon this. All negotiations with the Xhosa turned on one point, namely, that they had to abandon their claims to the Zuurveld and never again come west of the Fish with their livestock.

During the following years the Zuurveld remained the bone of contention, and in 1794 Adriaan van Jaarsveld declared in the heemraad meeting at Graaff-Reinet that he "was of the notion that to obtain a steadfast peace with the Xhosa it would have been best to give back to them the Zuurveld that had been their own land in former times."[106]

By the 1790s, however, a new factor made its appearance in the conflict between white and Bantu on the eastern frontier. The Xhosa (principally the vagrant class) began to steal the frontier farmers' livestock and this created an additional cause for friction. Gradually this factor increased in importance in such a way that it altogether overshadowed the original cause of the conflict between white and Bantu, which obviously still continued to exist.

But we will leave it there. Here we only wanted to show that stock thefts originally did not play such an important role as they did later in the border troubles, and that conflicts of interest over grazing land were an extremely important cause for the first conflicts with the Bantu.

NOTES

1. C 2280, Original Ordinance Book, Swellengrebel, 18 December 1739.

2. C 103, Resolutions, 2 July 1737, p. 304 and 15 October 1737, p. 301; C 133, Resolutions, 1755, p. 463.

3. Diary of Beutler's journey in G. M. Theal, *Belangrijke Historische Dokumenten over Zuid-Afrika* (Cape Town, 1896, 1911), II:16.

4. Journal of van Plettenberg's Journey, in Theal, *Belangrijke Historische Dokumenten*, I:25; Diary of Beutler's journey, Ibid., II:50; AR, Accession 1900, XXI, no. 85, Franz von Winkelman, Nachrichten der Östlichen Kaffern, 1789, p. 14.

5. CO 4438, Collins Papers, 1808–1809, Supplement, July 1809.

6. Diary of Beutler's Journey, in Theal, *Belangrijke Historische Dokumenten*, II:34, 38.

7. C 134, Resolutions, 11 May 1756, p. 246.

8. Journal of van Plettenberg's Journey, in Theal, *Belangrijke Historische Dokumenten*, I:23.

9. Ibid., p. 23.

10. Ibid., p. 26.

11. Swellengrebel Family Archives, van Plettenberg to Swellengrebel, 1 February 1779.

12. G. M. Theal, *History and Ethnography of South Africa before 1795* (London, 1909–1910), III:109.

13. Ibid., p. 107–09.

14. C 2683 Report of van Plettenberg regarding the Burgher Petition, 1782, p. 30.

15. Journal of van Plettenberg's Journey, in Theal, *Belangrijke Historische Dokumenten*, I:27.

16. H. Lichtenstein, *Reisen im südlichen Africa in den Jahren 1803, 1804, 1805 und 1806* (Berlin, 1811–1812), I:375.

17. Swellengrebel Family Archives, van Plettenberg to Swellengrebel, 7 March 1780; van Jaarsveld to Landdrost, 8 April 1779, in D. Moodie, *The Record; or a series of official papers relative to the condition and treatment of the native tribes in South Africa* (Cape Town, 1838), III:89.

18. C 2217, Logbook, 17 December 1779, p. 322.

19. 1/SWM 3/14, Testimonies, Stephanus Scheepers, 18 December 1779.

20. Ibid., H. C. Janse van Rensburg, 15 December 1779.

21. C 547, Letters Rec., van Ryneveld to Governor, 25 August 1779, p. 818.

22. 1/SWM 10/2, Logbook (Swellendam), 16 August 1779.

23. Ibid., 25 October 1779.

24. C 547, Letters Rec., van Ryneveld to Governor, 28 October 1779.

25. Ibid., Enclosure Potgieter to Landdrost, 27 September 1779, with van Ryneveld to Governor, 28 October 1779.

26. 1/SWM 3/14, Testimonies, Jan Scholts, 5 November 1779.

27. Ibid., J. H. Potgieter, 1 December 1779; Jan Scholts, 5 November 1779; S. Scheepers, 18 December 1779.

28. A search was made everywhere for these letters, but regrettably it was impossible to find them.

29. C 548, Letters Rec., De Wet to Governor, 13 March 1780.

30. C 2230, Logbook, Testimony of P. H. Ferairre before the Defence Council of Swellendam, 25 October 1780, p. 604ff.

31. Swellengrebel Family Archives, van Plettenberg to Swellengrebel, 12 May 1780; C 158, Resolutions, 25 July 1780, p. 228.

32. Testimony of A. C. Greyling, March 1836, in Moodie, *Record*, III:112.

33. C 2230, Logbook (Swellendam), 25 October 1780, p. 607.

34. C 548, Letters Rec., De Wet to Governor, 13 March 1780.

35. Theal, *History and Ethnography of South Africa before 1795*, III:128.

36. 1/STB 20/2, Letters Rec., van Jaarsveld, 22 June 1780. The letter containing the description of Bouwer's commando was not found.

37. 1/STB 10/162, Militia officer's reports about San robberies, Adriaan van Jaarsveld (Camdeboo), 4 October 1784.

38. Charges that Adriaan van Jaarsveld brought against militia officer Bothma are referred to in a defense council meeting at Swellendam on 3 December 1782. The charges are the same as those in the above-mentioned document and are also formulated in nearly the same words. This strengthens the supposition that Adriaan van Jaarsveld was the author of the document in question. And if that is the case, he thus must have drafted it between 6 August and 3 December 1782.

39. C 563, Letters Rec., Woeke et al. to Governor van der Graaf, 4 November 1786, p. 706; 1/GR 1/1, Resolutions, 13 November 1786, p. 5, paragraph 20.

40. C 563, Letters Rec., Woeke to Governor, November 1786, p. 685.

41. C 548, Letters Rec., De Wet to Governor, 13 March 1780.

42. C 158, Resolutions, 14 November 1780, p. 358.

43. 1/STB 20/2, Letters Rec., van Jaarsveld to Stellenbosch Landdrost, 22 June 1780.

44. Ibid.

45. C 158, Resolutions, 24 October 1780, p. 333. The appointment was made on this date.

46. C 2230, Logbook (Swellendam), 25 October 1780, p. 609ff.

47. C 549, Letters Rec., no. 84, A. van Jaarsveld to Landdrost, 11 September 1780.

48. C 158, Resolutions, 14 November 1780, p. 359:

It is thought proper and accordingly resolved, to fix the afore-mentioned Fish River as the boundary between our inhabitants and the Caffers, and to make this known in the Instructions for the Field Commandants in the far-lying Districts mentioned, with orders to the same, that in the event the frequently mentioned Caffers do not willingly honor their promises made to the Honorable Governor, and make their way to the other side of the Fish River, they must then be compelled with force to do so.

49. 1/STB 19/23, Diverse Instructions, Landdrost, Heemraden and Military Officers to Adriaan van Jaarsveld, 6 March 1781, paragraph 4.

50. Ibid., paragraph 10.

51. 1/STB 1/143, Judicial Documents and Reports, General Report of A. van Jaarsveld to Landdrost, Heemraden and Military Officers, 20 July 1781.

52. 1/STB 19/23, Diverse Instructions, 6 March 1781, paragraph 5.

53. Ibid., paragraph 10.

54. Ibid., paragraph 11.

55. 1/STB 1/143, Judicial Documents and Reports, 20 July 1781.

56. Ibid.

57. C 1275, Petitions, Reports, etc., no. 60, Landdrost, Heemraden and Military Officers to Governor, 2 October 1781.

58. C 160, Resolutions, 18 October 1781, p. 596.

59. Ibid., 6 November 1781, p. 640.

60. 1/STB 13/13, Diverse Reports and Letters to Landdrost and Military Council, A. van Jaarsveld to D. van Ryneveld and Military Council, 8 May 1782.

61. 1/STB 10/162, Militia Officers' reports, A. van Jaarsveld, 4 October 1784.

62. 1/SWM, Logbook, 25 October 1784; 1/GR 1/1, Resolutions, 3 December 1787, pp. 24–25.

63. C 1275, Petitions, Reports, etc., Instructions of 21 November 1781. [Trans. note: In the text and footnote of the Afrikaans edition this date is given as 1783. The date 1781 is correct.]

64. C 169, Resolutions, 28 August 1785, p. 601.

65. C 2288, Original Ordinance Book, part V, 19 July 1786, pp. 482–85.

66. C 170, Resolutions, 17 January 1786, p. 65; C 172, Resolutions, 19 July 1786, p. 858.

67. 1/STB 10/162, Militia Officers' reports, van Jaarsveld, 4 October 1784.

68. C 2288, Original Ordinance Book, part V, 19 June 1786.

69. C 563, Letters Rec., Woeke to Governor, November 1786, pp. 683–84; 8 December 1786.

70. C 565, Letters Rec., Woeke to Governor, 2 January 1787, p. 37.

71. C 563, Letters Rec., Woeke to Governor, 14 November 1786, p. 663.

72. 1/GR 8/1, Letters Rec., Governor to Woeke, 9 January 1787.

73. C 570, Letters Rec., Wägener to Governor, 9 October 1788.

74. Ibid.

75. C 570, Letters Rec., Enclosure of Wägener to Pieter de Buys and eight others, 3 October 1788, with letter of 9 October 1788 to the governor.

76. C 578, Letters Rec., Woeke to Governor, 9 February 1789, pp. 39–43.

77. Dr. Theal presents the events as follows:

[in] March, 1789, . . . a large body of Xosas, headed by the chiefs Langa, Cungwa, and others of less note, suddenly crossed the Fish River, and spread over the Zuurveld. The farmers fled before the invaders, but were unable to save the whole of their cattle. The landdrost immediately instructed the burgher captain Daniel Willem Kühne to take measures for the defence of the district, and despatched an express to Cape Town with a report that war was unavoidable and a request that the Council would send a hundred soldiers to his assistance. [*History and Ethnography of South Africa before 1795*, III:178]

Dr. Gie writes:

As a result of this [mutual altercations in Kafferland], and also because the farmers' cattle continued to lure them there, a large number of Xhosas burst into the Suurveld in March 1789. . . . Woeke had to wait for permission from Cape Town to use the commandos against the enemy. In the meantime the farmers, spread thinly over a large area, had to flee before the invaders, and again they had to look on as their houses were burned and their cattle stolen. [*Geskiedenis van Suid-Afrika*, I:217–18]

78. Woeke's report of 9 February 1789 was discussed on 20 March in the Council of Policy (see C 182, Resolutions, 1789, p. 367), and on 3 March 1789 Woeke reported that he had received word from Kühne that he had carried out his commission. (C 578, Letters Rec., p. 65)

79. C 578, Letters Rec., Woeke to Governor, 3 March 1789, pp. 65–66.

80. AR, Accession 1900, XXII, no. 85, Franz Von Winkelman, Nachrichten der Östlichen Kaffern, 1789, p. 14.

Dr. Theal gives the following version of the events:

In the meantime Captain Kühne had raised a commando, which was no sooner in the field than the invaders, without waiting to be attacked, fell back to the Fish River, which they found in flood, so that they were unable to cross. They were lying on the bank, and the burgher commando was approaching, when the instructions of the Council were received by the landdrost. He put on record his opinion that a fatal mistake to the future tranquility of the country was being made, and forwarded to the Council a letter to that effect; but carried out his orders with as much zeal as if they had been in accordance with his own views. The commando was at once discharged. Not a shot had been fired in retaliation for the losses sustained by the invasion, nor a single head of cattle recovered, so the burghers were indignant and almost mutinous when they were required to disband. [*History and Ethnography of South Africa before 1795*, III:179]

Dr. Gie presents these matters in the same light as Dr. Theal:

When Maynier arrived on the eastern border, a commando, which Woeke was compelled by necessity to call out, had the Xhosas trapped alongside the flooded banks of the Fish River. At last the farmers again had the opportunity to punish severely the destroyers of their farms, but the instructions from the Cape forbade an attack on the innocent natives and the commando had to be disbanded. The burghers turned back, bitterly disappointed and deeply shaken in their loyalty to the government. [*Geskiedenis van Suid-Africa*, I:218]

81. C 578, Letters Rec., Woeke to Governor, 3 March 1789, p. 66.

82. C 182, Resolutions, 20 March 1789, p. 367ff; 1/GR 8/1, Letters Rec., Governor to Woeke, 20 March 1789; 1/GR 1/1, Resolutions, 13 May 1789, pp. 94–96.

83. 1/GR 1/1, Resolutions, 13 May 1789, p. 94.

84. 1/GR 1/1, Resolutions, 13 May 1789, p. 95.

85. The commission's interesting report to the governor, dated 5 August 1789, is found in C 1301, Petitions, Reports, etc. We will primarily refer to this document in the following pages except where other sources are indicated.

86. Dr. Theal renders the question and response as follows:

The landdrost asked why they had invaded the Colony. They replied that they did not regard their action as an invasion, because they considered the country between the Fish River and the Kowie their own, as they had purchased it some years before from a Hottentot named Ruiter. [*History and Ethnography of South Africa before 1795*, III:180]

87. C 1301, Petitions, Reports, etc., enclosure no. 1, Petrus Lindeque to Barend Lindeque, 25 June 1789, with Report of the Commission (i.e., Landdrost Woeke of Graaf-Reinet et al.), 5 August 1789, new pp. 285–86.

88. C 1301, Petitions, Reports, etc., Report of the Commission, 5 August 1789.

89. C 1301, Petitions, Reports, etc., enclosure no. 2, Instructions to Wägener, 27 June 1789, with Report of the Commission, 5 August 1789.

90. C 1301, Petitions, Reports, etc., enclosure no. 4, L. Meyer to Wägener, 3 July 1789, with Report of J.J.F. Wägener, 5 July 1789.

91. C 1301, Petitions, Reports, etc., Report of J.J.F. Wägener, 5 July 1789.

92. C 1301, Petitions, Reports, etc., enclosure no. 2, Viljoen, Campfer, and Janse to Wägener, 3 July 1789, with Report of J.J.F. Wägener, 5 July 1789.

93. C 1301, Petitions, Reports, etc., enclosure no. 5, Wägener to Lucas Meyer, 4 July 1789, with Report of the Commission, 5 August 1789. See also Report of J.J.F. Wägener, 5 July 1789.

94. C 1301, Petitions, Reports, etc., enclosure no. A, Wägener to Delport, 5 July 1789, with Report of J.J.F. Wägener, 3 August 1789.

95. C 1301, Petitions, Reports, etc., Report of J.J.F. Wägener, 5 July 1789.

96. C 1301, Petitions, Reports, etc., enclosure no. 4, Landdrost and Heemraden to Wägener, 13 July 1789, with Report of J.J.F. Wägener.

97. C 1301, Petitions, Reports, etc., Report of J.J.F. Wägener, 3 August 1789.

98. C 1301, Petitions, Reports, etc., Wägener to Landdrost, 19 July 1789.

99. C 1301, Petitions, Reports, etc., Report of J.J.F. Wägener, 3 August 1789.

Dr. Theal also relates the arrangement that Wägener made and then continues:

with his report to this effect to the board of landdrost and heemraden of Graaff-Reinet on the 3rd of August, 1789, ends the official account of the Second Kaffir invasion of the Colony and of its disastrous consequences to the farmers. [*History and Ethnography of South Africa before 1795*, III:181]

100. C 1301, Petitions, Reports, etc., Report of the Commission, 5 August 1789.

101. 1/GR 1/9, Minutes and Annexures of the Defense Council meeting, Lucas Meyer to Woeke, 17 August 1789.

102. Ibid., Gerrit Scheepers, Sr., to Woeke, 27 December 1789.

103. Dr. Theal describes conditions on the frontier at this time as follows:

> In the documents issued by the government the arrangement made by Mr. Wägener was henceforth termed a restoration of peace and quietness; but the burghers of Graaff-Reinet chose to call it by a very different name. Almost any other people in the world would have abandoned a district exposed to such ruinous losses as those sustained by the hardy and perservering frontier Colonists at this period. But they were determined to hold their own. In the neighbourhood of the Xosas they guarded their herds with arms in their hands. While along the great mountain range they were continually struggling with the Bushman. [*History and Ethnography of South Africa before 1795*, III:181–82]

Dr. Gie writes:

> In August 1789 'peace' was then restored on this basis and a chronic condition of unrest and disorder on the frontier received the official sanction of the government. . . . The murder of whites and theft of their possessions increased. [*Geskiedenis van Suid-Afrika*, I:220]

104. See Theal, *History and Ethnography of South Africa before 1795*, III:177–82.

105. See the short letter of Gerrit Scheepers. On a later occasion J. A. Hurter, who also had to make an attempt to evacuate the Xhosa from the Zuurveld, pointed out to Tshaka that the farmers had to pay rent for the land that the Xhosa had allowed to be overgrazed. (1/GR 1/9, Minutes and Annexures of the Defense Council meeting, 15 February 1792, Journal of Hurter's meetings.)

106. 1/GR 1/1, Resolutions, 26 May 1794, p. 249.

Abbreviations

AR	Algemeen Rijksarchief [State Archives], The Hague
BO	(First) British Occupation archives, CAD
BR	Bataafse Republiek archives, CAD
C	Raad van Politic archives, CAD
CAD	Cape Archives Depot, Cape Town
CJ	Raad van Justie archives, CAD
CO	Colonial Office archives, CAD
CO 48/13	Colonial Office series 48/13, PRO, London
GH	Government House archives, CAD
GR	Graaff-Reinet archives, CAD
Kol. Arch.	Koloniaal Archief, AR, The Hague
PRO	Public Record Office, London
PSB	Preußische Staatsbibliothek, Berlin
RLR	Receiver of Land Revenue archives, CAD
SG	Old CAD series of Surveyor General; now RLR
St.	vdM abbreviation, Stellenbosch archives, CAD
STB	Present abbreviation, Stellenbosch archives, CAD
Sw.	vdM abbreviation, Swellendam archives, CAD
SWM	Present abbreviation, Swellendam archives, CAD
T	vdM abbreviation, Tulbagh archives, CAD
VC	Verbatim Copy, CAD
vdM	P. J. van der Merwe
VRS	Van Riebeeck Society
WOC	Present abbreviation, Worcester archives, CAD

Bibliography

Note to the Reader: Because of space limitations, not all sources cited in the text have been included in this bibliography. In such cases the specific bibliographical citation is given in the footnote. Refer as well to the Bibliography in my *Noordwaarstse Beweging van die Boere voor die Groot Trek*.

[Trans. note: This "Note to the Reader" appears at the end of the Bibliography in the Afrikaans edition. For the reader's convenience, I have included in the bibliography of this English edition all sources cited in the text.]

MODERN BOOKS AND ARTICLES

Macmillan, W. M. *The South African Agrarian Problem and Its Historical Development*. Johannesburg, 1919.

———. *Bantu, Boer, and Briton*. London, 1929.

Malan, J. H. *Boer en Barbaar; of, Die Lotgevalle van die Voortrekkers viral tussen die jare 1835–1840*. Potchefstroom, 1913.

———. *Boer en Barbaar; of, Die Geskiedenis van die Voortrekkers tussen die jare 1835–1840, en verder, van die Kaffernasies met wie hulle in aanraking gekom het*. 2d expanded ed. Bloemfontein, 1918.

Moffat, J. S. *The Lives of Robert and Mary Moffat*. London, 1890.

Moodie, D. "Saxon Nomads." In *South African Annals*. Pietermaritzburg, 1855.

———. *A Voice from Kahlamba*. South African Annals, no. 11. Pietermaritzburg, 1857.

Roux, P. E. *Die Verdedigingstelsel aan die Kaap onder die Hollands-Oosindiese Kompanjie*. Stellenbosch, 1925.

Theal, G. M. *History and Ethnography of South Africa before 1795*. 3 vols. London, 1909–10.

———. *History of South Africa since 1795*. 5 vols. London, 1907–08.

Van der Walt, A. J. H. *Die Ausdehnung der Kolonie am Kap der Guten Hoffnung, 1700–1779*. Berlin, 1928.

Watermeyer, E. B. *Three Lectures on the Cape of Good Hope under the Government of the Dutch East India Company.* Cape Town, 1857.

Wilmot, A., and J. C. Chase. *History of the Colony of the Cape of Good Hope from Its Discovery to the Year 1819.* Cape Town, 1869.

Walker, E. A. *Frontier Tradition in South Africa.* Oxford, 1930.

———. "Relief and the European Settlement of South Africa." *The Scottish Geographical Magazine* 46(1) (1930): 1–9.

CONTEMPORARY SOURCES

Contemporary Books and Articles

Albertyn, J. R. *Die Armblanke en die Maatskappy.* Stellenbosch, 1932. [English title: *The Poor White and Society.* Stellenbosch, 1932.]

Alexander, Sir James Edward. *Excursions in Western Africa, and Narrative of a Campaign in Kaffir-Land on the Staff of the Commander-in-Chief.* 2 vols. London, 1840.

Allamand, J. N. S., J. C. Klockner, and H. Hop. *Nieuwste en Beknopste Beschrijving van de Kaap der Goede Hoop.* Amsterdam, 1778.

Arbousset, J., and F. Daumas. *Narrative of an Exploratory Tour to the North-East of the Colony of the Cape of Good Hope.* Translated by J. C. Brown. London, 1852.

Backhouse, James. *A Narrative of a Visit to the Mauritius and South Africa.* London, 1844.

Barrington, George. *An Account of a Voyage to New South Wales.* 2 vols. London, 1810.

Barrow, J. *Travels into the Interior of Southern Africa.* London, 1806.

Baynes, C. R. *Notes and Reflections during a Ramble in the East, etc.* London, 1843.

Bird, W. W. *State of the Cape of Good Hope in 1822.* London, 1823. [Reprint. Cape Town: C. Struik, 1966.]

[Blount, Edward.] *Notes on the Cape of Good Hope made during an Excursion in that Colony in the year 1820.* London, 1821.

Bogaerts, A. *Historische Reizen door d' Oostersche Deelen van Azie.* Amsterdam, 1711.

Borcherds, P. B. *An Auto-Biographical Memoir.* Cape Town, 1861.

Boyce, William B. *Notes on South African Affairs, from 1834 to 1838, with Reference to the Civil, Political and Religious Condition of the Colonists and Aborigines.* Grahamstown, 1838.

Broadbent, S. *A Narrative of the First Introduction of Christianity Amongst the Barolong Tribe of Bechuanas, South Africa: With a Brief Summary of the Subsequent History of the Wesleyan Mission to the Same People.* London, 1865.

Bunbury, C. J. F. *Journal of a Residence at the Cape of Good Hope, with Excursions into the Interior.* London, 1848. [Reprint. New York: Negro Universities Press, 1969.]

Burchell, W. J. *Hints on Emigration to the Cape of Good Hope.* London, 1820.

———. *Travels in the Interior of Southern Africa.* 2 vols. London, 1822. [Reprint. London: Batchworth Press, 1953.]

Campbell, J. *Travels in Southern Africa.* London, 1815. [Reprint. Africana Collectanea Series, no. 47. Cape Town: C. Struik, 1974.]

———. *Travels in Southern Africa.* 2 vols. London, 1822. [Reprint (2 vols. in 1). New York: Johnson Reprint Corporation, 1967.]

Chase, J. C. *The Cape of Good Hope and the Eastern Province of Algoa Bay.* London, 1843.

Cleveland, R. J. *In the Forecastle, or Twenty-five Years a Sailor.* New York [ca. 1842].

Colenbrander, H. T. *De Afkomst der Boeren.* [The Hague], 1902. [Reprint. Cape Town: C. Struik, 1964.]

Collins, Col. R. *Aanteekeningen bij Gelengenheid eener Gouvernements Commissie naar de Zuidelijke Takken van de Rivier T'kij en door het Kafferland in het jaar 1809.* Cape Town, 1835. [Trans. note: See Papers of Col. R. Collins, relative to his journeys to the Northern and Eastern Districts of the Cape Colony, 1808–1809, in Cape Archives Depot, Cape Town, CO 4438, c.]

Cook, Captain. *A Voyage to the Pacific Ocean.* Vol. I. London, 1784.

[Cramer, Bastiaan.] *Trap der Jeugd, of Naauwkeurige en benattelyke grondlegger der Nederduitsche Spel-, Lees-, Schryf-, en Taalkunst.* Pietermaritzburg: P. Davis en Zonen, 1873.

Cumming, G. R. *Five years of a Hunters Life in the Far Interior of South Africa.* 2 vols. London, 1855.

[Curtis, C. G.] *An Account of the Colony of the Cape of Good Hope, with a view to the Information of Emigrants.* London, 1819.

Dapper, Olfert. *Nauwkeurige Beschrijvingen der Africanische Gewesten.* Amsterdam, 1668.

De Jong, C. *Reizen naar de Kaap de Goede Hoop.* 2 vols. Harlem, 1802–1803.

De Mist, Augusta Uitenhage. "Dagverhaal van een Reis naar de Kaap de Goede Hoop en in de Binnenlanden van Afrika door Jonkvr. Augusta Uitenhage de Mist in 1802 en 1803." *Penélopé* (Amsterdam), VIII(1835), pp. 72–127.

De Mist, J. A. *The Memorandum of Commissary J. A. de Mist.* Cape Town: Van Riebeeck Society, 1920. Series I, vol. 3.

Dryson, A. W. *Tales at the Outspan or Adventures in the Wild Regions of Southern Africa.* London, 1865.

Du Plessis, A. J. "Die Geskiedenis van die Graankultuur in Suid-Afrika: Tydens die Eerste Eeu, 1652–1752." M.A. thesis, University of Stellenbosch, 1929. [Trans. note: The page numbers given by vdM are from the thesis manuscript. This work was later published, under the same title, in the series *Annale van die Universiteit van Stellenbosch*, XI(Sept. 1933): 9–127, Cape Town: Nasionale Pers, 1933.]

Ebner, Johan Leonard. *Reise nach Süd-Afrika und Darstellung meiner während acht Jahren als Missionair unter den Hottentotten gemachten Erfahrungen, u.z.w.* Berlin, 1829.

Fawcett, John. *Account of an Eighteen Months' Residence in the Cape of Good Hope in 1835–36.* Cape Town, 1836.

Forbes, James. *Oriental Memoirs.* 4 vols. London, 1813.

Foster, G. *A Voyage round the World.* London, 1777.

Fouché, L. *Die Evolutie van die Trekboer: Lesing gehou voor die Christelike Jongelieden Vereniging.* Pretoria, 1909.

"Four Months in Cape Colony." *Chambers' Miscellany of Useful and Entertaining Tracts,* XX(173) (1847): 1–32.

Fruin, Robert. *Tien Jaren uit den Tachtigenjarigen Oorlog, 1588–1598.* The Hague, 1906.

Geyer, A. L. *Das Wirtschaftliche System der Niederländischen Ostindischen Kompanie am Kap der Guten Hoffnung.* Munich and Berlin, 1923.

Gie, S. F. N. *Geskiedenis van Suid-Afrika of Ons Verlede.* 2 vols. Stellenbosch, 1924.

Gleanings in Africa (by an English officer). London, 1806. [Reprint. New York: Negro Universities Press, 1969.]

Godlonton, R. *A Narrative of the Irruption of the Kafir Hordes into the Eastern Province of the Cape of Good Hope 1834–1835.* Grahamstown, 1836. [Reprint. Africana Collectanea Series, no. XI. Cape Town: C. Struik, 1965.] [Trans. note: In his bibliography, vdM gives the title of this book as *A Narrative of the Interruption* . . . and cites C. A. Fairbridge as the author. He uses this citation in his footnotes and repeats the citation in *Noordwaartse Beweging*. I can find no book with such a title, or any author named Fairbridge writing in South Africa at the time. Although vdM does not mention the Godlonton volume at all, as the page numbers to the references in the text match the Godlonton volume, I am certain that this is the correct citation. I have noted such in the relevant footnotes.]

———. *Sketches of the Eastern Districts of the Cape of Good Hope as they are in 1842.* Grahamstown, 1842.

Grosskopf, J. F. W. *Plattelandsverarming en Plaasverlating.* Stellenbosch,

1932. [English title: *Rural Impoverishment and Rural Exodus*. Stellenbosch, 1932.]

Harris, W. C. *Sketch of the Emigration of the Border Colonists.* [Full title: *Narrative of an Expedition in southern Africa during the years 1836, and 1837, from the Cape of Good Hope, through the territories of the chief Moselekatse, to the tropic of Capricorn, with a sketch of the Recent Emigration of the border Colonists, and a zoological appendix.*] [Bombay: at the American Mission Press, 1838.] [Published the next year in London as *The Wild Sports of Southern Africa.*]

————. *The Wild Sports of Southern Africa.* London, 1839. [Reprint of 1852 ed. Cape Town: C. Struik, 1963.]

Holman, J. *A Voyage round the World including Travels in Africa, etc. from 1827–1832.* London, 1834–35.

Howison, J. *European Colonies in various Parts of the World.* Vol. I. London, 1834.

Kay, Stephen. *Travels and Researches in Caffraria.* London, 1833.

Kolbe, P. *Nauwkeurige en Uitvoerigne Beschrijving van de Kaap de Goede Hoop.* Amsterdam, 1726.

Latrobe, C. *Journal of a Visit to South Africa in 1815 and 1816 with some account of the Missionary Settlements of the United Brethren near the Cape of Good Hope.* London, 1812. [Reprint of 1818 New York ed. New York: Negro Universities Press, 1969.]

Le Vaillant, F. *New Travels into the Interior Parts of Africa in the years 1783, 1784 and 1785.* 3 vols. London, 1786.

————. *Travels into the Interior Parts of Africa by the way of the Cape of Good Hope in the years 1780, 1781, 1782, 1783 and 1784.* 2 vols. London, 1790.

Lichtenstein, H. *Reisen im südlichen Africa in den Jahren 1803, 1804, 1805 und 1806.* 2 vols. Berlin, 1811–12. [Cape Town: Van Riebeeck Society, 1928–30, Series I, vols. 10, 11.]

Lockyer, C. *An Account of the Trade in India.* London, 1711.

MacCrone, I. D. *Race Attitudes in South Africa: Historical, Experimental and Psychological Studies.* London, 1937.

[MacGilchrist, J.] *The Cape of Good Hope (By a Traveller).* Glasgow, Edinburgh, London, 1844. 2d imp., London: Frank Cass, 1965.

Mackenzie, W. *Sketches of Travels in Southern Africa.* Edinburgh, 1824.

Malherbe, E. G. *Onderwys en die Armblanke.* Stellenbosch, 1932. [English title: *Education and the Poor White Problem.* Stellenbosch, 1932.]

Masson, F. "An Account of Three Journeys from Cape Town into the Southern Parts of Africa." *Philosophical Transactions of the Royal Society.* London, 1776, 66:268–318.

Mentzel, O. F. *Vollständige und Zuverläszige Geographische und Topographische*

Beschreibung des Berühmten und in aller Betrachtung merkwürdigen Afrikanischen Vorgebirges der Guten Hoffnung. 2 vols. Glogau, 1785, 1787.

———. *A Complete and Authentic Geographical and Topographical Description of the Cape of Good Hope* (Translation of Parts I and II). Cape Town: Van Riebeeck Society, 1921, 1924, series I, vols. 4, 6. [See also Cape Town: Van Riebeeck Society, 1944, series I, vol. 25.]

Methuen, H. H. *Life in the Wilderness; or, Wanderings in South Africa*. London, 1848.

Moffat, R. *Journey from Colesberg to Steinkopf, 1831–1835*. Read before the Royal Geographical Society of London on the 14th of December 1857. *Journal of the Royal Geographical Society*. (London, 1858) 28:153–73. [Reprint. London: W. Clowes and Sons [1858.]

Moodie, J. W. D. *Ten Years in Southern Africa*. London, 1835.

Napier, Lt. Col. E. E. *Excursions in Southern Africa, including a History of the Cape Colony, an Account of the Native Tribes, etc.* 2 vols. London, 1849.

Nicholson, George. *The Cape and its Colonists with Hints to Settlers*. London, 1848.

Nightingale, Anna Elizabeth. *Gleanings from the South, East and West*. London, 1843.

Ovington, J. *A Voyage to Surat in the year 1689*. London, 1696.

Paravicini di Capelli, W. B. C. *Reize in de Binnenlanden van Zuid-Afrika gedaan in den jaare 1803*. (Ms. in the Koloniale Biblioteek, The Hague.) [Cape Town: Van Riebeeck Society, 1965, series I, vol. 46.]

[Parlby, Fanny Parks.] *Wanderings of a Pilgrim, in search of the picturesque, during four and twenty years in the East*. Vol. II. London: Pelham Richardson, 1850.

Paterson, Wm. *A Narrative of Four Journeys into the Country of the Hottentots and Caffraria in the Years 1777, 1778 and 1779*. London, 1779.

Percival, R. *An Account of the Cape of Good Hope*. London, 1804. [Reprint. New York: Negro Universities Press, 1969.]

Philip, J. *Researches in South Africa*. 2 vols. London, 1828.

[Philipps, T.] *Scenes and Occurrences in Albany and Cafferland, South Africa*. London, 1827.

Polson, Nicolas. *Subaltern's Sick Leave; or, Rough Notes on a Visit in Search of Health to China and the Cape of Good Hope*. Calcutta, 1837.

Pringle, T. *African Sketches*. London, 1834.

Prior, James. *Voyage in the Indian Seas, to the Cape of Good Hope, etc., during the Years 1810 and 1811*. London, 1820.

Raikes, H. *Memoirs of the Life and Services of Vice-Admiral Sir Jahleel Brenton, Baronet, K.C.B.* London, 1846.

Renshaw, R. *Voyage to the Cape of Good Hope and up the Red Sea; with Travels in Egypt, through the Deserts, etc., in the course of the last war.* Manchester, 1804.

Reynolds, J. N. *Voyage of the United States Frigate Potomac under the command of Commodore John Downes during the Circumnavigation of the Globe in the years 1831, 1832, 1833 and 1834.* New York, 1835.

Robertson, G. A. *Notes on Africa—to which is added an appendix, containing a Compendius Account of the Cape of Good Hope, its Productions and Resources with a variety of important information very necessary to be known by persons about to emigrate to that Colony.* London, 1819.

Rose, Cowper. *Four Years in Southern Africa.* London, 1829.

Saint Pierre, J. H. Bernardin de. *Voyage to the Island of Mauritius (or Isle-de-France), the Isle of Bourbon, the Cape of Good Hope, etc.* Translated by J. Parish. London, 1775.

Shaw, Barnabas. *Memorials of South Africa.* 2d ed. London, 1841. [Reprint. Cape Town: C. Struik, 1966.]

Shipp, J. *Memories of the extraordinary military career of John Shipp, late a Lieut. in His Majesty's 87th Regiment.* London, 1890.

Smith, Dr. A. *Report of the Expedition for Exploring Central Africa from the Cape of Good Hope, 28th June, 1834.* Cape Town, 1836.

Smith, T. *South Africa Delineated.* London, 1850.

Sparrman, A. *Reize naar de Kaap de Goede Hoop, de Landen van den Zuid Pool, en rondom de Waereld.* 2 vols. Leiden, 1786. [Cape Town: Van Riebeeck Society, 1975, 1977, series II, vols. 6, 7.]

Spoelstra, C. *Bouwstoffen voor de geschiedenis der Nederduitsch-Gereformeerde Kerken in Zuid-Afrika.* 2 vols. Amsterdam, 1906.

Stavorinus, J. S. *Voyages to the East-Indies.* 3 vols. London, 1798.

Steedman, A. *Wanderings and Adventures in the Interior of Southern Africa.* 2 vols. London, 1835.

Stout, B. *Narrative of the Loss of the Ship Hercules Commanded by Captain Benjamin Stout, on the Coast of Caffraria the 16th of June, 1796; also a Circumstantial Detail of his Travels through the Southern Deserts of Africa and the Colonies of the Cape of Good Hope.* London, 1798.

———. *The Cape of Good Hope and its Dependencies.* London, 1820.

Swellengrebel, Hendrik, Jr. *Briefwisseling van Hendrik Swellengrebel, Jr. oor Kaapse Sake, 1778–1792.* Edited by G. J. Schutte. English summary by A. J. Böeseken. Assisted by H. M. Robertson. Cape Town: Van Riebeeck Society, 1982, series II, vol. 13. [Trans. note: See also *Swellengrebel Archives* below.]

Teenstra, M. D. *De Vruchten mijner Werkzaamheden gedurende mijne Reize over de Kaap de Goede Hoop, naar Java en terug, over St. Helena, naar de Nederlanden.* Groningen, 1830. [Cape Town: Van Riebeeck Society, 1943, series I, vol. 24.]

Theunissen, J. B. N. *Aantekeningen van een reis door de Binnenlanden van Zuid-Afrika van Port Elizabeth naar de Kaapstad gedaan in 1823.* Oostende, 1824.

Thompson, G. *Travels and Adventures in Southern Africa.* 2 vols. London, 1827. [Cape Town: Van Riebeeck Society, 1967, 1968, series I, vols. 48, 49.]

Thunberg, K. P. *Travels in Europe, Africa and Asia performed within the years 1770 and 1779.* 2 vols. London, 1793. [Cape Town: Van Riebeeck Society, 1986, series II, vol 17.]

A Travellar, *The Cape of Good Hope: A Review of its Present Position as a Colony: Information which may be of advantage to an Intending Settler.* Glasgow, Edinburgh, and London, 1844.

Valentyn, F. *Oud en Nieuw Oost Indiën.* Part V. Amsterdam, 1727. [Cape Town: Van Riebeeck Society, 1971, 1973, series II, vols. 2, 4.]

Van der Merwe, P. J. *Die Noordwaartse Beweging van die Boere voor die Groot Trek (1770–1842).* The Hague, 1937. [Reprint. Pretoria: State Library, 1988.]

————. *TREK. Studies oor die Mobiliteit van die Pioniersbevolking aan die Kaap.* Cape Town, 1945.

Van Hoogendorp, G. K. *Verhandelingen over den Oost-Indischen Handel.* 2 vols. Amsterdam, 1801.

Van Pallandt, Baron Andries. *General Remarks on the Cape of Good Hope,* Cape Town, 1803. [Reprint. Cape Town: Printed for the Trustees of the South African Public Library, 1917.] [Trans. note: Translated from the French. De Mist ordered all copies of the work to be destroyed. Only one copy of the original, in French, survives, and it is in the Cape Archives.]

Van Reenen, D. G. *Dagverhaal eener Reize naar de Binnenlanden van Afrika beoosten de Kaap de Goede Hoop geleegen, in den Jaare 1803, gedaan door Z.E. den gouverneur en General en Chief, J. W. Janssens, etc.* (Ms. in the Cape Archives). [Cape Town: Van Riebeeck Society, 1938, series I, vol. 18.]

Van Reenen, J. *Journal of a Journey from the Cape of Good Hope undertaken in 1790 and 1791 by Jacob van Reenen in Search of the wreck of the Grosvenor.* Translated by E. Riou. London, 1792. [Cape Town: Van Riebeeck Society, 1927, series I, vol. 8.]

Von Meyer, W. *Reisen in Süd-Afrika wärend der Jahre 1840 und 1841.* Hamburg, 1843.

Wakefield, E. G. *A View on the Art of Colonization.* London, 1849.

Wilcocks, R. W. *Die Armblanke.* Stellenbosch, 1932. [English title: *The Poor White.* Stellenbosch, 1932.]

Weyermüller, F. *Die Völker Süd-Afrika's nach Augenzeugen geschildert und die Geschichte des Mussetse.* Strasbourg, 1842.
[Wynne, Frederick.] *Sketches of India (together with notes on the Cape of Good Hope).* London, 1816.

Contemporary Newspapers

Grahamstown Journal (to 1842).
Zuid-Afrikaansche Tijdschrift (to 1842).

ARCHIVAL SOURCES

Published Sources

Unofficial Sources

Collectanea. Cape Town: Van Riebeeck Society, 1924, series I, vol. 5.
Fouché, L. *Het Dagboek van Adam Tas, 1705–1706.* London, 1914. [Cape Town: Van Riebeeck Society, 1970, series II, vol. 1.]
Godée Molsbergen, E.C. *Reizen in Zuid-Afrika.* 4 vols. The Hague, 1932.
Moodie, D. *The Record; or, a series of official papers relative to the condition and treatment of the native tribes in South Africa, I, III and V.* Cape Town, 1838. [Reprint. Amsterdam: A. A. Balkema, 1960.]
Preller, G. S. *Voortrekkermense.* 4 vols. Cape Town, 1920–25.
Reports of De Chavonnes and His Council, and of Van Imhoff, on the Cape. Cape Town: Van Riebeeck Society, 1919, series I, vol. 1.
Theal, G. M. *Belangrijke Historische Dokumenten over Zuid-Afrika.* 3 vols. Cape Town, 1896, 1911.
———. *Chronicles of the Cape Commanders.* Cape Town, 1882.
———. *Records of the Cape Colony, 1793–1831.* 36 vols. Cape Town, 1897–1905.
Van der Heiden, J., and Adam Tas. "Contra-Deductie. . . ." In *The Defence of William Adriaan van der Stel.* (*Precis of the Archives of the Cape of Good Hope*). Edited by H. C. V. Leibbrandt. Cape Town: W. H. Richards & Sons, 1897.
Van der Stel, W. A. "Korte Deductie. . . ." In *The Defence of William Adriaan van der Stel.* (*Precis of the Archives of the Cape of Good Hope*). Edited by H. C. V. Leibbrandt. Cape Town: W. H. Richards & Sons, 1897.

Bluebooks and other Government Publications

Bigge, J. I., Colebrooke, W. M. G., and Blair, W. *Reports of the Commissioners of Inquiry on the Cape of Good Hope:* (1) *Upon the Administration of the Government at the Cape of Good Hope.* (2) *Upon the Finances at the Cape of Good Hope (Dated 6 September 1826).* Cape Town, 1827.

Harding, Walter. *The Cape of Good Hope Government Proclamations from 1806–1825 and Ordinances passed in Council 1825–1844.* 3 vols. Cape Town, 1839–45.

Ordonantie Raakende het Bestier der Buiten Districten in de Nederlandsche Zuid-Africaansche Volksplanting aan de Kaap de Goede Hoop. Cape Town, 1805.

Unpublished Sources

Cape Archives Depot, Cape Town

Stellenbosch Archive [Trans. note: Now located under 1/STB]

Document no.		Title	Dates
St	1–21	Resolutiën (Landdrost en Heemraden)	1691–1795
	122	Inkomende Brieven	1732–1748
	124	Inkomende Brieven aan Landdrost	1759–1785
	125	Inkomende Brieven aan Landdrost	1763–1794
	135	Particuliere Inkomende Brieven	1795–1798
	205	Uitgaande Brieven	1784–1789
	206	Uitgaande Brieven	1789–1793
	207	Uitgaande Brieven en Memoriën	1795–1796
	228	Minuut Brieven	1716–1746
	229	Minuut Afgaande Brieven	1759–1846
	230	Minuut Brieven	1766–1785
	328	Bilietten	1686–1822
	330	Bilietten van Landdrost en Heemraden	1729–1793
	338	Placaten, Advertentiën, etc.	1686–1783
	339	Placaten, Advertentiën, etc.	1740–1824
	340	Placaten, Advertentiën, etc.	1725–1794
	341	Proclamations, etc.	1795–1806
	346	Rapporten van Gecommitteerde Heemraden	1757–1784
	347	Inkomende Brieven (Rapporten)	1775–1795
	348	Rapporten der door Landdrost en Heemraden gedane Commissiën	1785–1793
	491–95	Relaasen en Verklaringen (sommige	

		Krimineel)	1740–1796
676–88		Verklaringen en Visitatien	1693–1795
691–701		Krijgsraad Resolutiën	1693–1795
703		Regelment voor Krijgsraad van Stellenbosch	1790
719		Diverse Rapporten en Brieven aan Landdrost en Krijgsraaden	1780–1797
725		Diverse Instructiën	1686–1811
726		Commando Papieren	1776–1846
728		Opgaaf Lijsten (van diensdoende Burgers) en Brieven	1776–1778
732		Veldwagtmeesters Rapporten over Bosjesmans Roverijen	1773–1793
733		Veldwachtmeesters Rapporten	1773–1795
786		Leenings Plaatsen Versoeken	1667–1796
787		Correspondentie re Leenings Plaatsen	1793–1795
793		Lijst van Nieuwe en Verlatene Leenings Plaatsen	1794
797		Diverse Stukken re Leenings Plaatsen	1770–1830
975		Summaria der Opgaven [Tax rolls]	1741–1809
1116–17		Veelijst van Stellenbosch en Drakenstein	1783–1796
1131		Over Bergsche Saken	1744–1745

Raad van Politic (C)

C	1–109	Resolutiën	1655–1795
	291–325	Memoriën en Rapporten	1710–1791
	409–77	Inkomende Brieven	1752–1795
	493–573	Uitgaande Brieven	1652–1795
	650–57	Dag Register van Stellenbosch en Drakenstein en Swellendam	1729–1784
	668	Berigt door H.A. Wikar nopens zijne Omzwervingen langs de Grote Rivier	1778
	680–87	Origineel Placcaat Boek, Dl. 1–8	1652–1795
	700–07	Memoriën en Instructiën	1657–1795
	710	Instructiën hier gelaten door Goew. Gen. Van Imhoff met Bijlagen	1743
	735	Rapport van G. van Plettenberg over Eene Memorie der Burgers	1781–1782
		Attestatiën	
		Bijlagen	
		Dag Register (v.d. Kaap)	

Raad van Justie (CJ)

CJ   1596–1604 Crimineele en Civiele Regs. Rolle   1652–1727

Graaff-Reinet Archive [Trans. note: Now located under 1/GR]
GR   55   Inkomende Brieven   1785–1803
      328   Generale Opgaaf [Tax rolls]   1788–1825
      395   Staat en Constitutie der Colonie   1787–1801
      434–44   Judicieële Verklaringen   1786–1826
      445   Notariële Verklaringen   1787–1826
      581   Resolutiën van Landdrost en
                 Heemraden   1786–1795
      585   Notulen en Bijlagen van Krijgsraad
                 Vergad.   1787–1795
      595   Bijlagen en Notulen van
                 gecommitteerden
      637   Burgher Lists
GR Ongekatalogiseer [Uncatalogued in the 1930s]
Fieldcornets, Veldwagtmesters and Veldkommandant: vol. 1, 1787–1809
Letters Despatched by H. C. D. Maynier, Commissioner Graaff-Reinet
Letters Received at Graaff-Reinet from other Private Individuals and
   Field Cornets
Private Individuals   vol. I 1801–1845
                                                   vol. II 1781–1800

Swellendam Archive [Trans. note: Now located under 1/SWM]
Sw   9   Notulen van Landdrost en Heemraden
                 (Defect)   1747, 1750, 1751
Sw   10   Notulen van Landdrost en Heemraden
                 (Defect)   1773–1785
Sw   11   Notulen van Landdrost en Heemraden
                 (Defect)   1789–1798
Sw   12   Concept Resolutiën: Landdrost en
                 Heemraden   1789–1804
Sw   13   Extract Resolusiën van de Raad van
                 Politie   1792–1795
Sw   15   Dag Register
                           1755–1776, 1777, 1779, 1780, 1782, 1786
Sw   44–48   Inkomende Brieven   1744–1795
Sw   123   Verklaringen   1739
Sw   124–29   Verklaringen zoo Civiele als Crimineele 1763–1795
Sw   130–31   Attestatiën

Sw	150–55	Citatiën	1749–1795
Sw	184	Proclamations	1790–1793
Sw	202	Opgaaf Rollen [Tax rolls]	1793
Sw	437	Vendurollen	1793

Office of the Surveyor General [Trans. note: Old SG is now located under Receiver of Land Revenue (RLR)]

SG	1–38	Oude Wildschutte Boeken (Old Gamehunters Books)	1687–1793
	39–41	Generaal Alphabeth van de namen der Landbouwers en Namen der veeplaatsen	
	42	Alphabetische Naamlijst van Lenings Plaatsen en Bezitters	1794–1798
	45–47	Aantekeningboek	1775–1791
	50	Verzameling van Placcaten en Resolutiën met die Instructies betreffende 's Lands Generale Inkomsten	1686–1794

Varia

VC (Verbatim Copy)   76   H. A. Campagne, memorie en bijzonderheden wegens de Overgave der Kaap de Goede Hoop 1795 (Get. 17 Mei 1797)

J   83–143   Opgaaflijste vir Graaf-Reinet [Tax rolls]   1787–1840

[Trans. note: There are several citations in the footnotes from the CO, BO, BR, GH, and WOC series in the Cape Archives and the CO series in the Public Record Office, London. Van der Merwe did not include those series in the bibliography of the Afrikaans edition. I include them here with the most recent volume numbers and titles.]

Colonial Office (CO)

CO	41	Court of Justice	1812
CO	49	Court of Justice	1813
CO	53	Surveyor of Lands	1813
CO	60	Court of Justice	1814
CO	64	Inspector of Lands and Woods	1814
CO	79	Inspector of Lands	1816
CO	173	Collector of Tithes and Land Revenue	1822
CO	197	Inspector of Lands and Woods	1823

CO	336	Commissioner General	1828
CO	2567	Letters Received from Graaff Reinet	1809
CO	2568	Letters Received from Tulbagh	1809
CO	2580	Letters Received from Graaff-Reinet	1812
CO	2614	Letters Received from Tulbagh	1818
CO	2695	Letters Received from Graaff-Reinet	1827
CO	2696	Letters Received from Worcester	1827
CO	2744	Letters Received from Worcester	1833
CO	4438	Collins Papers, 1808–1809, Supplement July	1809
CO	4439	Correspondence re the distribution of lands and the rights of land owners to certain waste lands	1810–1811
CO	4443	Correspondence relative to the allegations made by the missionaries Read and van der Kemp	1811
CO	4835	Letters Dispatched 24 October 1813 to 4 April 1814	

First British Occupation (BO)

| BO | 49–55 | Letters from Stellenbosch | 1796–1803 |

Bataafse Republiek (BR)

BR	131	Scheeps and other Journals	
BR	390	Ordinance concerning the Supervision of the Interior Districts	
BR	444	Minutes of Commissioner-General de Mist, Account of a Conference on Agriculture, 31 October 1803	
BR	504	Original Ordinance Book: Extract Resolutions	

Government House (GH)

| GH | 23/8 | General Dispatches from Secretary of State, London, 7 March 1826 to 16 December | 1826 |

Worcester Archive (1/WOC)

| WOC | 1/6 | Resolutions | 1827 |

Colonial Office, Public Record Office, London (CO)

| CO | 48/13 | Secretary of State, Original Correspondence | 1811 |

Algemeen Rijksarchief [State Archives], The Hague (AR)

Aanwinsten (Accession) 1914 (Kolleksie van J. van Plettenberg)

No. 26a   Journaal van een Reisje gedaan naar de Saldanha en St.
               Helena Baayen, 1776.
No. 27   Dagregister van een Tocht in de Kaap door Pieter
               Cloeten, 10 September to 26 December 1776.
No. 27a   Journaal van een Reisje, gedaan naar het Heeren
               Logements Gebergte en de mond der Olifants Rivier in
               het Noorden der Colonie, 1777.
No. 28   Dagregister van den Tocht tot aan de Plettenberg's en
               Groote Rivier door Olof Gotlief de Wet, 1778.
No. 29–30   Reisverhalen van zijn Tochten langs de Groote Rivier
               door Hendrik Jacob Wikar, 1778–1779, (2 parts)
        Aanwinsten 1890 (Kolleksie van Beelaerts van Blokland)
no. 30 (m)   Zaak van H. C. D. Maynier, 1802.
        Aanwinsten 1900
no. XX   Journaal en Verhaal eener Land Reijze in den Jaar 1803
               door J. W. Janssens.
        Aanwinsten (Accession) (Nederburg Kolleksie)
Nos. 40, 157, 168, 169, 242, 548a
        Koloniaal Archief
Nos. 3969–4338   Brieven en Papieren van Cabo de Bonne Esperance
                   overgekomen (1657–1794). [Trans. note: Each new
                   number for this set of documents is higher by 22
                   than the corresponding old numbers. Thus, 3969
                   is now 3991, 3970 is now 3992, etc.]
No. 4464 [new no. 4918]   Verantwoording van den Hr. Gouverneur te
                   Cabo de Goede Hoop, C.J. v.d. Graaff,
                   met bijlaag.
No. 4384 [new no. 4924]   Reflectiën op de antwoorden van C. J. v.d.
                   Graaff met bijlaag.
No. 4384 [new no. 4924]   Van Renen, Artoys, Roos en Heyns:
                   Memorie, 7 Mei 1779.
        Raad der Aziatische Bezittingen en Etablissementen
No. 292   Missives van den Commissaris-Generaal over de Kaap
               de Goede Hoop, Mr. J.A. de Mist, aan den Raad
               met Bijlagen, 1802–1805
Nos. 293–303   Notulen en Handelingen van Commissaris-Generaal
               over de Kaap de Goede Hoop, Mr. J. A. de Mist
               met Bijlagen, 1802–1805.
Nos. 309–10   Missives van de Gouverneur en Raad van de Kaap de
               Goede Hoop aan den Raad met Bijlagen, 1803–1806.
No. 311   Missives van den Gouverneur en Generaal de Chev.
               aan de Kaap de Goede Hoop aan den Raad met

Bijlagen, 1803–1806.

No. 350   Missives van bijzondere Kaapsche Ambtenaren aan den
Raad, met Bijlagen, (8 October to 12 September
1803).

Swellengrebel Archives

[Trans. note: VdM did not include any reference to the Swellengrebel
Archives in the bibliography of the Afrikaans edition. The reader is referred to the volume of selected documents from the Swellengrebel Archives published in 1982 by the Van Riebeeck Society, *Briefwisseling van
Hendrik Swellengrebel, Jr. oor Kaapse Sake, 1778–1792*, that I have included
in the list of contemporary books above. On page 41 in the English introduction to this volume, G. J. Schutte states that the archives are currently
in the possession of Mr. N. J. A. C. Swellengrebel, retired sea captain in
Hilversum, and descendant of one of Hendrik Swellengrebel's brothers.]

Preußische Staatsbibliothek (PSB)

[Trans. note: Since unification, this archive is known as the Staatsbibliothek zu Berlin, Preußsischer Kulturbesitz.]

   Mss. Germ. Octav. (M. H. K. Lichtenstein)
No. 275   Mist, J. A. de: Memoranda van de Kaapsche Landreize
1803 (9 October 1803 to 23 March 1804). [Trans. note:
This ms. is incorrectly listed as no. 257 in the
Afrikaans edition.]

   Mss. Germ. Quarto
No. 857   Berichte des General-Gouverneurs der Kap-Kolonie, J. W.
Janssens an den General-Commissar J(acob) A(braham)
de Mist über die inneren Verhältnisse der Kolonie,
1804–1805 (aus der Kanzlei des de Mist).

   Mss. Germ. Fol.
No. 879   Reize uit Nederland naar de Kaap de Goede Hoop en
terug. En voorts het Dagverhaal van een Landreize in
de Binnenlanden op de Zuidhoek van Afrika gedaan
door J(acob) A(braham) de Mist in de Jaaren 1802,
1803, 1804 en 1805, gevolgd door een beschrijving en
beschouwing van den tegenwoordigen toestand dier
Volkplanting en derzelven nabuurige natiën.

No. 880   Journaal en Verhaal eener Landreise in den Jaare 1803
door den gouverneur en Generaal deezer Colonie J. W.
Janssens door de Binnenlanden van Zuijd Africa
gedaan.

No. 881   Kladjournaal der Reize gedaan door den Commissaris-

Generaal J(acob) A(braham) de Mist en deszelfs gevolg in de Binnenlanden van Afrika, 1803–1804.

No. 883   Notul of Dagverhaal der Reis en verrichtingen van Praesident en Gecommitterde Leeden uit de Commissie van Veeteelt en Landbou in de beiden Roggevelden, den Hantam, etc. (1805).

No. 884   Instruktionen des General-Kommissars Jacob Abraham de Mist betr. das Justizwesen, Steuer und Finanzwesen, Vormundschaftsangelegenheiten, etc. der Kap Kolonie, 1803–1804.

No. 885   Instructionen und Memoires für die Befehlshaber der Kap Kolonie von 1802. Zusammengestelltes Aktenstück aus der Kanzlei von Jacob Abraham de Mist. Darin: Schriftwechsel zwischen J. A. de Mist und J. W. Janssens. Briefwechsel aus den Jahren 1792–1793 über den Verteidigungs Zustand der Kolonie (Abschriften Holland).

No. 887   Zusammengestellte Schriftstücke und Notizen aus der Kanzlei des General-Kommissars Jacob Abraham de Mist, enthaltend vergleichende Statist. Materialen betr. das Finanzwesen, Bevölkerungs Fragen, Kirchen-und Schulsachen auch Landbau, Naturgeschichte und Viehzucht der Kap Kolonie der Jahre 1795–1804. Holland.

No. 888   Berichte des General-Gouverneurs der Kap Kolonie J. W. Janssens an den General-Kommissar Jacob Abraham de Mist betr. die inneren Verhältnisse der Kolonie, besonders die Eingeborenen-Fragen, nebst Beilagen (darunter Verhandlungs Protokolle mit den Kaffer-Kapitänen) 1803–1804. Holland.

No. 889   Concept-Gouvernement voor de Kaap de Goede Hoop (Entworfen Febr., 1801).

No. 890   Lichtenstein, Martin Heinrich Karl: Geschichte der Cap Colonie (Materialen und Entwürfe); Geschichte der Entdeckung und Colonisation des Südlichen Afrika (Autograph).

No. 892   Een Generale Beschrijving van de Kolonie de Kaap de Goede Hoop en de vier distrikten, waar uit dezelve is zamengesteld, 1798.

No. 893   Memorie over de Caab de Goede Hoop, aan het Gouvernement der Fransche Republiek gepresenteerd.

No. 894   Memorie over de Kaap de Goede Hoop (Ueber den

          Wirtschaft Wert des Landes) I en II.
No. 895   Memorie over de Kaap de Goede Hoop III.
No. 896   Schriftwechsel zwischen dem Gouverneur der Kap
          Kolonie J. W. Janssens und dem General-Kommissar
          Jacob Abraham de Mist über Verwaltungsangelegenheiten,
          1804–1805 (Originale) Holland.
No. 897   Schrift Wechsel zwischen dem General Gouverneur der
          Kap Kolonie J. W. Janssens und dem General-
          Kommissar Jacob Abraham de Mist in den Politischen
          Angelegenheiten der Jahre 1803–1805, nebst Schreiben
          der Batavischen Republik (Geheimakten aus der Kanzlei
          des J. A. de Mist). Holland.

[Trans. note: The citations used by Dr. van der Merwe for the sources
in Berlin are from:
Degering, Hermann. *Kurzes Verzeichnis der Germanischen Handschriften der
    Preussischen Staatsbibliothek. I. Die Handschriften in Folioformat. II. Die
    Handschriften in Quartformat. III. Die Handschriften in Oktavformat und
    Register zu Band I-III.* Leipzig: Karl W. Hiersemann Verlag, 1925,
    1926, 1932.]

Index

275